Geometric Modeling and Algebraic Geometry

Bert Jüttler • Ragni Piene
Editors

Geometric Modeling
and Algebraic Geometry

 Springer

Bert Jüttler

Institute of Applied Geometry
Johannes Kepler University
Altenberger Str. 69
4040 Linz, Austria
bert.juettler@jku.at

Ragni Piene

CMA and Department of Mathematics
University of Oslo
P.O.Box 1053 Blindern
0136 Oslo, Norway
ragnip@math.uio.no

ISBN: 978-3-540-72184-0   e-ISBN: 978-3-540-72185-7

Library of Congress Control Number: 2007935446

Mathematics Subject Classification Numbers (2000): 65D17, 68U06, 53A05, 14P05, 14J26

© Springer-Verlag Berlin Heidelberg 2008

Cover design: WMX Design GmbH, Heidelberg

Printed on acid-free paper

9 8 7 6 5 4 3 2 1

springer.com

Preface

The two fields of Geometric Modeling and Algebraic Geometry, though closely related, are traditionally represented by two almost disjoint scientific communities. Both fields deal with objects defined by algebraic equations, but the objects are studied in different ways. While algebraic geometry has developed impressive results for understanding the theoretical nature of these objects, geometric modeling focuses on practical applications of virtual shapes defined by algebraic equations. Recently, however, interaction between the two fields has stimulated new research. For instance, algorithms for solving intersection problems have benefited from contributions from the algebraic side.

The workshop series on Algebraic Geometry and Geometric Modeling (Vilnius 2002[1], Nice 2004[2]) and on Computational Methods for Algebraic Spline Surfaces (Kefermarkt 2003[3], Oslo 2005) have provided a forum for the interaction between the two fields. The present volume presents revised papers which have grown out of the 2005 Oslo workshop, which was aligned with the final review of the European project GAIA II, entitled *Intersection algorithms for geometry based IT-applications using approximate algebraic methods* (IST 2001-35512)[4].

It consists of 12 chapters, which are organized in 3 parts. The first part describes the aims and the results of the GAIA II project. Part 2 consists of 5 chapters covering results about special algebraic surfaces, such as Steiner surfaces, surfaces with many real singularities, monoid hypersurfaces, canal surfaces, and tensor-product surfaces of bidegree (1,2). The third part describes various algorithms for geometric computing. This includes chapters on parameterization, computation and analysis of ridges and umbilical points, surface-surface intersections, topology analysis and approximate implicitization.

---

[1] R. Goldman and R. Krasauskas, *Topics in Algebraic Geometry and Geometric Modeling*, Contemporary Mathematics, American Mathematical Society 2003.

[2] M. Elkadi, B. Mourrain and R. Piene, *Algebraic Geometry and Geometric Modeling*, Springer 2006.

[3] T. Dokken and B. Jüttler, *Computational Methods for Algebraic Spline Surfaces*, Springer 2005.

[4] http://www.sintef.no/IST_GAIA

The editors are indebted to the reviewers, whose comments have helped greatly to identify the manuscripts suitable for publication, and for improving many of them substantially. Thanks to Springer for the constructive cooperation during the production of this book. Special thanks go to Ms. Bayer for compiling the LaTeX sources into a single coherent manuscript.

Oslo and Linz,   *Bert Jüttler*
August 2007   *Ragni Piene*

Contents

Part I Survey of the European project GAIA II

1 The GAIA Project on Intersection and Implicitization
T. Dokken 5

Part II Some special algebraic surfaces

2 Some Covariants Related to Steiner Surfaces
F. Aries, E. Briand, C. Bruchou 31

**3 Real Line Arrangements
and Surfaces with Many Real Nodes**
S. Breske, O. Labs, D. van Straten 47

4 Monoid Hypersurfaces
P. H. Johansen, M. Løberg, R. Piene 55

5 Canal Surfaces Defined by Quadratic Families of Spheres
R. Krasauskas, S. Zube 79

6 General Classification of (1,2) Parametric Surfaces in $\mathbb{P}^3$
T.-H. Lê, A. Galligo 93

Part III Algorithms for geometric computing

**7 Curve Parametrization over Optimal Field Extensions
Exploiting the Newton Polygon**
T. Beck, J. Schicho 119

8 Ridges and Umbilics of Polynomial Parametric Surfaces
F. Cazals, J.-C. Faugère, M. Pouget, F. Rouillier 141

9 Intersecting Biquadratic Bézier Surface Patches
S. Chau, M. Oberneder, A. Galligo, B. Jüttler 161

10 Cube Decompositions by Eigenvectors of Quadratic Multivariate Splines
I. Ivrissimtzis, H.-P. Seidel 181

11 Subdivision Methods for the Topology of 2d and 3d Implicit Curves
C. Liang, B. Mourrain, J.-P. Pavone 199

12 Approximate Implicitization of Space Curves and of Surfaces of Revolution
M. Shalaby, B. Jüttler 215

Index 229

Part I

Survey of the European project GAIA II

The European project GAIA II entitled *Intersection algorithms for geometry based IT-applications using approximate algebraic methods* (IST 2001-35512) involved six academic and industrial partners from five countries. The project aimed at combining knowledge from Computer Aided Geometric Design, classical algebraic geometry and real symbolic computation in order to improve intersection algorithms for Computer Aided Design systems. The project has has produced more than 50 scientific publications and several software toolkits, which are now partly available under the GNU GPL license.

We invited the coordinator of the project, Tor Dokken, to present a survey describing the background, the methods, the results and the achievements of the GAIA project. His summary is the first part of this volume.

1

The GAIA Project on Intersection and Implicitization

Tor Dokken

SINTEF ICT, Department of Mathematics
tor.dokken@sintef.no

Summary. In the GAIA-project we have combined knowledge from Computer Aided Geometric Design (CAGD), classical algebraic geometry and real symbolic computing to improve intersection algorithms for Computer Aided Design (CAD) systems. The focus has been on:

- Singular and near singular intersections of surfaces, where the surfaces are parallel or near parallel along segments of the intersection curves.
- Self-intersection of surfaces to facilitate trimming of self-intersecting surfaces.

The project has published more than 50 papers. Software toolkits from the project are available for downloading under the GNU GPL license.

1.1 Introduction

In the GAIA project we have combined knowledge from Computer Aided Geometric Design (CAGD), classical algebraic geometry and real symbolic computing to improve intersection algorithms for CAD-type systems. The calculation of the intersection between curves or surfaces can seem mathematically simple. This is true for the intersection of e.g. two straight lines when they intersect transversally and closed expressions for finding the intersection are used. However, if floating point arithmetic is used, care has to be taken to properly handle situations when the lines are parallel or near parallel. The intersection of two bi-cubic parametric surfaces can be reduced to finding the real zero set of a polynomial equation $f(s,t) = 0$ of bi-degree (54,54), which by itself is a challenging problem. In industrial systems floating point arithmetic is used, thus introducing rounding errors. In CAD system there are tolerances defining when two points are to be regarded as the same point. This has also to be taken into consideration in CAD-related intersection algorithms. The consequence of low quality intersection algorithms in CAD-systems is low quality CAD-models. Low quality CAD-models impose high costs on the product creation processes in industry.

The objectives of the GAIA project were related both to the scientific and technological aspects:

- To establish the theoretical foundation for a new generation of methods for intersection and self-intersection calculation of 3D CAD-type sculptured surfaces by introducing approximate algebraic methods and qualitative geometric descriptions.
- To demonstrate through software prototypes the feasibility of the approach.
- To investigate other uses of the approximate algebraic methods developed for extending functionality in modeling and interrogation of 3D geometries.
- To demonstrate that cooperation between mathematical domains such as approximation theory, classical algebraic geometry and computer aided geometric design is an important part of improving mathematical based technology on computers.
- To interact with CAD systems developers to improve both friendly use and robustness of future CAD systems.

To address these objectives the project activities have been structured in four main work areas, where each partner has had one or two work areas as their main focus:

- **Classification**, where we have used the tools and knowledge of classical algebraic geometry to better understand singularities, see Section 1.5.
- **Implicitization**, where we have looked into resultants and approximate implicitization to better find exact and approximate implicit representations of parametric surfaces, see Section 1.6.
- **Intersection**, where we have looked into algebraic intersection methods, combined numeric and algebraic intersection algorithms, and combined recursive and approximate implicit intersection methods, see Section 1.7.
- **Applications**, where we have searched for other problem domains where the approach of approximate implicitization can be used for better solving challenging problems related to systems of polynomial equations, see Section 1.8.

In addition to the topics above we will in this paper also address:

- Project background and partners in Section 1.2.
- Why CAD-type intersection is still a challenge in industry in Section 1.3.
- The algorithmic challenges of CAD type intersections in Section 1.4.
- The potential impact of the GAIA project in section 1.9.

The list of references at the end of this paper is a bibliography of papers related to the GAIA-project published by the project partners during and after the GAIA-project.

1.2 Project background and facts

The Ph.D. dissertation *Aspects of Intersection Algorithms* [16] from 1997 established close dialogue between the Department of Applied Mathematics at SINTEF ICT in Oslo, and the algebraic geometry group in the Department of Mathematics, University of Oslo. Gradually the idea of establishing a closer cooperation with other European groups matured, and the algebraic geometry group at the University of Nice

Sophia Antipolis in France was contacted. An application for an IST FET Open Assessment project was made also including the CAD-company think3. The proposal was successful, and in October 2000 the project *IST 1999-290010 – GAIA – Application of approximate algebraic geometry in industrial computer aided geometry* was started.

The final review of the assessment project in October 2001 was successful, and the project consortium was invited to propose a full FET-Open Project. Also this proposal was successful, and July 1st 2002 the project *IST-2001-35512 – GAIA II – Intersection algorithms for geometry based IT-applications using approximate algebraic methods* started. The full project ended on September 30th 2005. The budgets of the phases of project have been:

- GAIA assessment phase: Budget: 175 000 EURO, with financial contribution from the European Union of 100 000 EURO.
- GAIA II project phase: Budget: 2 300 000 EURO, with financial contribution from the European Union of 1 500 000 EURO.

Among the project partners we find one CAD-company, one industrial research institute, and four university groups:

- **SINTEF ICT, Department of Applied Mathematics, Norway**, has been the project coordinator, and focused on work within approximate implicitization, recursive intersection algorithms and recursive self-intersection algorithms. For more information on SINTEF see: http://www.sintef.no/math/.
- **think3 SPA, Italy and France,** is a CAD-system developer, and had as their main role to supply industrial level examples of challenging CAD-intersection and self-intersections, to integrate developed intersection algorithms into a prototype version of their system thinkdesign, and finally to test and assess the prototype algorithms developed in the project. For more information on think3 see: http://www.think3.com/.
- **University of Nice Sophia Antipolis (UNSA), France,** developed in close cooperation with INRIA exact intersection algorithms and a triangulation based reference method for surface intersection and self-intersection. For more information on UNSA and INRIA see: http://www-sop.inria.fr/galaad/.
- **University of Cantabria, Spain,** worked on combined numeric and exact intersection algorithms. For more information see: http://www.unican.es/.
- **Johannes Kepler University, Austria,** focused on new approaches to approximate implicitization and testing of approximate implicitization algorithms. For more information on this partner see: http://www.ag.jku.at/.
- **University of Oslo, Norway,** has focused on classification of algebraic curves and surfaces and their singularities. For more information on the University of Oslo see: http://www.cma.uio.no/.

Based on state-of-the-art reports, research reports and software prototypes we have tried to establish a common mathematical understanding of different approaches and tools. As the project partners come from an axis spanning from fairly theoretical classical algebraic geometry to computer aided geometric design and CAD-system developers, a major focus has been on bridging the language and knowledge gaps between the different mathematical groups involved. All groups have had to invest time into better understanding the traditional approaches of the other groups.

1.3 Why are CAD-type intersections still a problem for industry?

1.3.1 CAD technology evolution hampered by standardization

In the Workshop on Mathematical Foundations of CAD (Mathematical Sciences Research Institute, Berkeley, CA. June 4-5, 1999) the consensus was that: *The single greatest cause of poor reliability of CAD systems is lack of topologically consistent surface intersection algorithms.* Tom Peters, Computer Science and Engineering, The University of Connecticut, estimated the cost to be $1 Billion/year. For more information consult SIAM News, Volume 32, Number 5, June 1999, *Closing the Gap Between CAD Model and Downstream Application*, http://www.siam.org/siamnews/06-99/cadmodel.htm. Too low quality of CAD-intersection forces the industry to resort to expensive workarounds and redesigns to develop new products.

CAD-systems play a central role in most producing industries. The investment in CAD-model representation of current industrial products is enormous. CAD-models are important in all stages of the product life-cycle, some products have a short lifetime, while other products are expected to last at least for one decade. Consequently backward compatibility of CAD-systems with respect to functionality and the ability to handle "old" CAD-models is extremely important to the industry. Transfer of CAD-models between systems from different CAD-system vendors is essential to support a flexible product creation value chain. In the late 1980s the development of the **STEP standard** (ISO 10303) *Product Data Representation and Exchange* started with the aim to support backward compatibility of CAD-models and CAD-model exchange. STEP is now an important component in all CAD-systems and has been an important component in the globalization of design and production. However, STEP standardized the geometry processing technology of the 1980s, and the problems associated with that generation of technology. Due to the CAD-model legacy (the huge bulk of existing CAD-models) upgraded CAD-technology has to handle existing models to protect the resources already invested in CAD-models. Consequently the CAD-customers and CAD-vendors are conservative, and new technology has to be backward compliant. Improved intersection algorithms have thus to be compliant with STEP representation of geometry and the traditional approach to CAD coming from the late 1980s. For research within CAD-type intersection algorithms to be of interest to producing industries and CAD-vendors backward compatibility and the legacy of existing CAD-models have not to be forgotten.

1.4 Challenges of CAD-type intersections

If the faces of a CAD-represented volume are all planar, then it is fairly straight-forward to represent the curves describing the edges with minimal rounding error.

However, if the faces are sculptured surfaces, e.g., bicubic NURBS - NonUniform Rational B-splines, the edges will in general be free form space curves with no simple closed mathematical description. As the tradition (and standard) within CAD is to represent such curves as NURBS curves, approximation of edge geometry with NURBS curves is necessary. For more information on the challenges of CAD-type intersections consult [54].

When designing within a CAD-system, *point equality tolerances* are defined that determine when two points should be regarded as the same. A typical value for such tolerances is 10^{-3}mm, however, some systems use tolerances as small as 10^{-6}mm. The smaller this tolerance is, the higher the quality of the CAD-model will be. Approximating the edge geometry with e.g., cubic spline interpolation that has fourth order convergence using a tolerance of 10^{-6} instead 10^{-3} will typically increase the amount of data necessary for representing the edge approximation by a factor between 5 and 6. Often the spatial extent of the CAD-models is around 1 meter. Using an approximation tolerance of 10^{-3}mm is thus an error of 10^{-6} relative to the spatial extent of the model.

The intersection functionality of a CAD-system must be able to recognise the topology of a model in the system. This implies that intersections between two faces that are limited by the same edge must be found. The complexity of finding an intersection depends on relative behaviour of the surfaces intersected along the intersection curve:

- **Transversal intersections** are intersection curves where the normals of the two surfaces intersected are well separated along the intersection curve. It is fairly simple to identify and localise the branches of the intersection when we only have transversal intersection.
- **Singular and near singular intersections** take place when the normals of the two surfaces intersected are parallel or near parallel in single points or along intervals of an intersection curve. In these cases the identification of the intersection branches is a major challenge.

Figures 1.1 and 1.2 respectively show transversal and near-singular intersection situations. In Figure 1.1 there is one unique intersection curve. The two surfaces in Figure 1.2 do not really intersect, there is a distance of 10^{-7} between the surfaces, but they are expected to be regarded as intersecting. To be able to find this curve, the point equality tolerance of the CAD-system must be considered. The intersection problem then becomes: Given two sculptured surface $f(u,v)$ and $g(s,t)$, find all points where $|f(u,v) - g(s,t)| < \varepsilon$ where ε is the point equality tolerance.

1.4.1 The algebraic complexity of intersections

The simplest example of an intersection of two curves in $\mathbb{R}^2$ is the intersection of two straight lines. Let two straight lines be given:

- A straight line represented as a parametric curve

Fig. 1.1. Transversal intersection between two sculptured surfaces

Fig. 1.2. Tangential intersection between two surfaces

$$\mathbf{p}(t) = \mathbf{P}_0 + t\mathbf{T}_0, t \in \mathbb{R},$$

with $\mathbf{P}_0$ a point on the line and $\mathbf{T}_0$ the tangent direction of the line.

- A straight line represented as an implicit equation

$$q(x,y) = ((x,y) - \mathbf{P}_1) \cdot \mathbf{N}_1 = 0, (x,y) \in \mathbb{R}^2,$$

with $\mathbf{P}_1$ a point on the line, and $\mathbf{N}_1$ the normal of the line.

Combining the parametric and implicit representation the intersection is described by $q(\mathbf{p}(t)) = 0$, a linear equation in the variable t. Using exact arithmetic it is easy to classify the solution as:

- An empty set, if the lines are parallel.
- The whole line, if the lines coincide.
- One point, if lines are non-parallel.

Next we look at the intersection of two rational parametric curves of degree n and d, respectively. From algebraic geometry it is known that a rational parametric curve of degree d is contained in an implicit parametric curve of total degree d, see [27].

- The first curve is described as a rational parametric curve

$$\mathbf{p}(t) = \frac{\mathbf{p}_n t^n + \mathbf{p}_{d-1} t^{n-1} + \ldots + \mathbf{p}_0}{h_n t^n + h_{n-1} t^{n-1} + \ldots + h_0}.$$

- The second curve is described as an implicit curve of total degree d

$$q(x, y) = \sum_{i=0}^{d} \sum_{j=0}^{d-i} c_{i,j} x^i y^j = 0.$$

By combining the parametric and implicit representations, the intersection is described by $q(\mathbf{p}(t)) = 0$. This is a degree $n \times d$ equation in the variable t. As even the general quintic equation cannot be solved algebraically, a closed expression for the zeros of $q(\mathbf{p}(t))$ can in general only be given for $n \times d \leq 4$. Thus, in general, the intersection of two rational cubic curves cannot be found as a closed expression. In CAD-systems we are not interested in the whole infinite curve, but only a bounded portion of the curve. So approaches and representations that can help us to limit the extent of the curves and the number of possible intersections will be advantageous.

We now turn to intersections of two surfaces. Let $\mathbf{p}(s, t)$ be a rational tensor product surface of bi-degree (n_1, n_2),

$$\mathbf{p}(s, t) = \frac{\displaystyle\sum_{i=0}^{n_1} \sum_{j=0}^{n_2} \mathbf{p}_{i,j} s^i t^j}{\displaystyle\sum_{i=0}^{n_1} \sum_{j=0}^{n_2} h_{i,j} s^i t^j}.$$

From algebraic geometry it is known that the implicit representation of $\mathbf{p}(s, t)$ has total algebraic degree $d = 2n_1 n_2$. The number of monomials in a polynomial of total degree d in 3 variables is $\binom{d+3}{3} = \frac{(d+1)(d+2)(d+3)}{6}$. So a bicubic rational surface has an implicit equation of total degree 18. This has 1330 monomials with corresponding coefficients.

Using this fact we can look at the complexity of the intersection of two rational bicubic surfaces $\mathbf{p}_1(u, v)$ and $\mathbf{p}_2(s, t)$. Assume that we know the implicit equation

$q_2(x, y, z) = 0$ of $\mathbf{p}_2(s, t)$. Combining the parametric description of $\mathbf{p}_1(u, v)$ and the implicit representation $q_2(x, y, z) = 0$ of $\mathbf{p}_2(s, t)$, we get $q_2(\mathbf{p}_1(u, v)) = 0$. This is a tensor product polynomial of bi-degree $(54, 54)$. The intersection of two bicubic patches is converted to finding the zero of

$$q_2(\mathbf{p}_1(u, v)) = \sum_{i=0}^{54} \sum_{j=0}^{54} c_{i,j} u^i v^j = 0.$$

This polynomial has $55 \times 55 = 3025$ monomials with corresponding coefficients, describing an algebraic curve of total degree 108. This illustrates that the intersection of seemingly simple surfaces can results in a very complex intersection topology. As in the case of curves, the surfaces we consider in CAGD are bounded, and we are interested in the solution only in a limited interval $(u, v) \in [a, b] \times [c, d]$.

1.5 Extend the use of algebraic geometry within CAD

The work within the GAIA project related to algebraic geometry and CAD has addressed three main topics:

- **Resultants** are one of the traditional methods for exact implicitization of rational parametric curves and surfaces. GAIA has produced some new results within this classical research area.
- **Singularities** in algebraic curves and surfaces are for understanding their geometry and topology.
- **Classification** is an old tradition in the field of Algebraic Geometry. It is a natural starting point when trying to understand the geometry of algebraic objects.

Papers on CAGD and algebraic methods from the project are [8, 9, 32, 33, 34, 35, 41, 42, 44, 48, 49, 57].

1.5.1 Resultants

The objective has been to develop tools for constructing, manipulating and exploiting implicit representations for parametric curves and surfaces based on resultant computations. The work in GAIA has been divided into three parts:

- A survey in four parts addressing:
  1. A resultant approach to detecting intersecting curves in P^3.
  2. Implicitizing rational hypersurfaces using approximation complexes.
  3. Using projection operators in Computer Aided Design.
  4. The method of moving surfaces for the implicitization of rational parametric surface in P^3.
- A report addressing sparse/toric resultant, results when the number of monomials is small compared to the number of possible monomials for polynomial of the degree in question.

- Development of prototypes of tools for constructing, manipulating and exploiting implicit representations for parametric curves and surfaces based on resultant computations.

One paper from the project addressing resultants is [7].

1.5.2 Singularities

Understanding the singularities of algebraic curves and surfaces is important for understanding the geometry of these curves and surfaces. A difficult problem in CAGD is the handling of self-intersections, and the theory of singularities of algebraic varieties is potentially a tool for handling this problem. In the GAIA project special emphasis has been put on detecting and locating singularities appearing on parameterized and implicitly given curves and surfaces of low degree.

- The presence of singularities affects the geometry of complex and real projective hypersurfaces and of their complements. We have illustrated the general principles and the main results by many explicit examples involving curves and surfaces.
- We have classified and analyzed the singularities of a surface patch given by a parameterization in order to proceed to an early detection. We distinguish algebraically defined surface patches and procedural surfaces given by evaluation of a program. Also we distinguish between singularities which can be detected by a local analysis of the parameterization and those which require a global analysis, and are more difficult to achieve.
- The detection of singularities is a critical ingredient of many geometrical problems, in particular in intersection operations. Once these critical points are located, one can for instance safely use numerical methods to follow curve branches. Detecting a singularity in a domain may also help in combining several types of methods.

A paper addressing singularities from the project is [48].

1.5.3 Classification

To use algebraic curves and surfaces in CAGD one needs to know about their shape: topology, singularities, self-intersections, etc. Most of this kind of classification theory is performed for algebraic curves and surfaces defined over the complex numbers, i.e., one considers complex (instead of only real) solutions to polynomial equations in two or three variables (or in three or four homogeneous variables, if the curves and surfaces are considered in projective space). Complete classification results exist only for low degree varieties (implicit curves and surfaces) and mostly only in the complex case. A simple example, the classification of conic sections, illustrates well that the classification over the real numbers is much more complicated than over the complex numbers.

We have collected known results about such classifications, especially concerning results for real curves and surfaces of low degree. Of particular interest in CAGD are parameterizable (i.e. so-called rational) curves and surfaces, and we have made explicit studies of various such objects. These objects, or patches of these objects, are potential candidates for approximate implicitization problems. For example, when the rough shape of a patch to be approximated is known, one can choose from a "catalogue" what kind of parameterized patch that is suitable - this eliminates many unknowns in the process of finding an equation for the approximating object and will therefore speed up the application. In addition to the survey of known results, particular objects that have been studied are:

- monoid curves and surfaces, especially quartic monoid surfaces
- tangent developables
- triangle and tensor surfaces of low degree of low (bi)degrees

Papers from the project addressing classification are [41, 42].

1.6 Exact and approximate implicitization

In CAD-type algorithms, combining parametric and algebraic representation of surfaces is in many algorithms advantageous. However, for surfaces of algebraic degree higher than two this is in general a very challenging task. E.g., a rational bi-cubic surface has algebraic degree 18. All rational surfaces have an algebraic representation. However, for surfaces of total degree higher than 3, not all algebraic surfaces will have a rational parametric representation. In the project we have the following two main approaches for change of representation.

1.6.1 Exact implicitization of rational parametric surfaces

General resultant techniques, but also specialized methods have been reviewed or developed in the GAIA II project to address the implicitization process:

- Projective, as well as anisotropic, resultants when the polynomials $f_0, \ldots, f_3$ have no base points.
- Residual resultants when the polynomials have base points which are known and have special properties.
- Determinants of the so-called approximation complexes which give an implicit equation of the image of the polynomials as soon as the base points are locally defined by at most two equations.

Papers from the project addressing topics of exact implicitization are [6, 23, 24, 27, 47].

Approach	Comment	Addressed in GAIA II
Triangulation	Will both miss branches and produce false branches	See section 1.7.1 on the Reference Method
Lattice evaluation	Will miss branches	Used in many CAD-systems. Not addressed in GAIA II
Recursive	Guarantees topology within specified tolerances	See section 1.7.2 addressing the combination of recursion and approximate implicitization
Exact	Guarantees topology however will not always work	The AXEL library see Section 1.7.3
Combined exact & numeric	Guarantees topology however will not always work, faster than the exact methods	Uses Sturm Harbicht sequences for topology of algebraic curves, see Section 1.7.4

Table 1.1. Different CAD-intersection methods and their properties.

1.6.2 Approximate implicitization of rational parametric surfaces

Two main approaches have been pursued in the project.

- **Approximate implicitization by factorization** is a numerically stable method that reformulates implicitization to finding the smaller singular values of a matrix of real numbers. See one of [17, 21] for an introduction. The approach can be used as an exact implicitization method if the proper degree is chosen for the unknown implicit and exact arithmetic is used. The approach has high convergence rates and is numerical stable. Strategies for selecting solutions with a desired gradient behavior are supplied, either for encouraging vanishing gradients or avoiding vanishing gradients. The approach works both for rational parametric curves and surfaces, and for procedural surfaces. Experiments with piecewise algebraic curves and surfaces have produced implicit curves and surfaces that have more vanishing gradients than is desirable. We have experienced that estimating gradients will improve this situation. We have established a connection between the original approach to approximate implicitization, and a numerical integration based method that can also be used for procedural surfaces, and a sampling/interpolation based approach [22].
- **Approximate implicitization by point sampling and normal estimates** is constructive in nature as it estimates gradients of the implicit representation to ensure that gradients do not vanish when not desired [1, 2, 3, 11, 13, 36, 37, 38, 40, 50, 51, 52]. The approach produce good implicit curves and surfaces and the problem of vanish gradients in not desired regions is minimal. The method works well for approximation by piecewise implicit curves and surfaces.

The work within GAIA has illustrated the feasibility of approximate implicitization, established both new methods on approximate implicitization with respect to theory and practical use of approximate implicitization. It has also been important to compare the different approaches to approximate implicitization [59].

1.7 Intersection algorithms

In the GAIA II project phase the work on the reference method, see 1.7.1, continued from the assessment phase was completed. Further a completely new recursive intersection code has been developed addressing industrial CAD-type problems. Two more research oriented intersection codes have been developed: A pure symbolic code and a combined symbolic numeric code. See Table 1.1 for a short overview.

1.7.1 The reference method

The reference method is based on intersecting triangulations that approximate surfaces. This can be used for getting a fast impression of the possible existence of intersection or self-intersections. However, as the approach is sampling based, there is no guarantee that all intersections are found, the triangulations intersected can easily produce an incorrect topology of the intersection in near singular and singular cases, and even false intersection branches might be found. The development of the reference method has been important to allow think3 to develop the new user interfaces, and experiment with these before the software from the combined recursive and approximate implicit intersection code was available in its first versions.

1.7.2 Combined recursive and approximate implicitization intersection method

The combined recursive and approximate implicitization intersection was an extremely ambitious implementation task, the challenges of the implementation and approach is discussed in [20]. The ambition has been to address the very complex singular and near singular intersections. The aim was also Open Source distribution. Consequently a completely new intersection kernel had to be developed to ensure that we do not have any copyright problems. A major challenge with respect to self-intersections is the complexity of cusp curves intersecting self-intersection curves. The traditional approaches for recursive subdivision based intersection algorithms do not work properly in these cases. Thus when starting to test the code we entered unknown territory. By the end of GAIA II we could demonstrate that the approach works, but the stability of the toolkit was not at an industrial level. However, stabilization work on the code has continued after the GAIA II project.

Recursive intersection codes traditionally use Sinha's theorem that states that for a closed intersection loop to exist in the intersection of two surfaces then the normal fields of the surfaces have to overlap inside the loop. Consequently if there is no overlap of the normal fields of two surfaces they can not intersect in a closed loop. However, in singular intersections normal fields will overlap. In near singular intersections even deep levels of subdivision often do not separate the normal fields. In the GAIA II program code we have used approximate implicitization for separating the spatial extent of the surfaces, and for analyzing the possibilities of closed intersection loops by combining an approximate implicitization of one surface with the parametric representation of the other surface. This is a very efficient tool when

NURBS surfaces approximating low degree algebraic surfaces are intersected. Such approximating NURBS surfaces are frequently bi-cubic and are thus much more challenging to intersect that the algebraic surfaces they approximate. Approximate implicitization is used to find the approximate algebraic degree of the surfaces, and consequently simplifies the intersection problem significantly.

The high-level reference documentation of the software has already been produced in doxygen and is available on the web. Other papers on numeric intersection algorithms from the project are [5, 14, 54, 55, 56].

1.7.3 Algebraic methods

The problems encountered in CAGD are sometimes reminiscent of 19th century problems. At that time, realizing the difficulties one had working in affine instead of projective space, and over the real numbers instead of the complex numbers one soon shifted the theoretical work towards projective geometry over the complex numbers. In fact, it is still in this situation that the modern intersection theory from algebraic geometry works best:

- **Bisection through a Multidimensional Sturm Theorem.** A variant of the classical Sturm sequence is presented for computing the number of real solutions of a polynomial system of equations inside a prescribed box. The advantage of this technique is based on the possibility of being used to derive bisection algorithms towards the isolation of the searched real solutions.
- **Algorithms for exact intersection.** Algorithms using Sturm–Habicht based methods have been implemented and are available at Axel - Algebraic Software Components for gEometric modeLing.

Papers on exact intersection methods from the project are [15, 30].

1.7.4 Combined algebraic numeric methods

The approach for the combined methods is to combine the rational parametric description of one surface $\mathbf{p}_1(s,t)$, with the algebraic representation of the other surface $q_2(x,y,z) = 0$. Thus the problem is converted to a problem of finding the topology of an algebraic curve $q_2(\mathbf{p}_1(s,t)) = 0$ in the parameterization of the first surface:

- **A limited number of critical points.** The approach is based on finding critical point, points where either $\nabla f(s,t) = 0$ or $\partial f(s,t)/\partial s = 0$. For any value in the first parameter direction of $f(s,t)$ there will be a limited number of such critical points. There is also a finite number of rotations of $f(s,t)$ that will have more than one critical point. $f(s,t)$ is rotated to ensure that for a given value there will be only one critical point.
- **Projection to first parameter direction.** The problem is project to a polynomial in the first parameter variable of $f(s,t)$ by computing the discriminant $R(s)$ of $f(s,t)$ with respect to t, and finding the real root of $R(s)$, $\alpha_1, \ldots, \alpha_r$.. The

Sturm-Habicht sequences here supply an exact number of real roots in the interval of interest.

- **Finding values in the second parameter direction.** Then for each α_i $i = 1, \ldots, r$ we compute the real roots of $f(\alpha_i, t)$, $\beta_{i,j}$, $j = 1, \ldots, s_i$. For every α_i and $\beta_{i,j}$ compute the number of half branches to the right and left of the point $(\alpha_i, \beta_{i,j})$.
- **Reconstruction of topology of the algebraic curve.** From the above information the topology of the algebraic curve in the domain of interest can be constructed.

Papers on this approach in the project are [4, 10, 28, 29, 31].

To ensure the approach to work the root computation has to use extended precision to ensure that we reproduce the number of roots predicted by the Sturm-Habicht sequences. The algorithms have been developed using symbolic packages.

1.8 New applications of the approach of approximate implicitization

A number of different applications of approximated implicitization are addressed in the subsections following.

1.8.1 Closest point foot point calculations

Inspired by approximate implicitization this problem has been addressed by modeling moving surfaces normal to the surface and intersecting in constant parameter lines [57]. The set up of the problems follows the ideas of approximate implicitization; singular value decomposition is used to find the coefficients of the moving surfaces. By inserting the coordinates of a point into such a moving surface a polynomial equation in one variable results. The zeros of this identify constant parameter lines with a foot point. Further a theory addressing the algebraic and parametric degree of the moving surface is established.

1.8.2 Constraint solving

Multiple constraints described by parametric curves, surfaces or hypersurfaces over a domain used for optimization can be modeled using approximate implicitization as a piecewise algebraic curve, or surface, or hypersurface. Thus a very compact way of modeling constraints has been identified.

1.8.3 Robotics

Within robotics we have identified a number of uses. We have experimented with checking for self-intersection of robot tracks. CAD-surfaces used in robot planning can check for self-intersections by the GAIA tools. The control of advanced robots can be expressed as systems of polynomial equations. To solve such equations the approaches of GAIA II for finding intersection and self-intersection e.g. using recursive subdivision and the Bernstein basis are natural extensions of the GAIA work. However, except for the exact methods developed, not much of the code generated in GAIA II can be directly used.

1.8.4 Micro and nano technology

We followed the suggestion by the reviewers at the second review (June 2004) to look at micro and nano technology and go to the DATE 2005 exhibition in Munich. Before this exhibition we tried to understand what the actual needs within nano and micro technology were. This proved to be a big challenge. Within SINTEF we both have a micro/nano technology laboratory and people doing ASIC design. First addressing those running the laboratory we realized that the laboratory was oriented towards production processes and could not answer our questions. Approaching ACIS designers was more successful. With the current level of circuit miniaturization, the actual geometry of the circuits due to etching starts to be more important. In the fine detail corners are not sharp, they are round. Thus to take the actual geometry of the circuits into consideration for simulation seems to be critical in micro and nano technology. During our presentation at the University boot of DATE we established two areas where the GAIA II approach can be used:

- Solution of systems of equations describing the properties of integrated circuits.
- Description of the detailed shape of circuits using piecewise algebraic surfaces.

However, within micro and nano technology there are already groups of mathematicians. To be able to address this area we have to establish a common meeting place, such as a series of workshops may be as a strategic support action in the 7th framework program.

1.9 Potential impact of the GAIA project

The development of mathematics for CAD has been stagnating since the standardization of CAD-representation in the start of the 1990s, and as the mathematicians addressing CAD-challenges got fewer. The CAD-vendors have merged to a handful of dominant world wide CAD-systems. As large user groups do not need handling of complex surface geometries, the problems of industries in need of improvements or improved algorithms have been given low priority by the vendors.

1.9.1 Bottleneck before GAIA II: Only rudimentary self-intersection algorithms

Advance shaped products are to a large extent built by structures of sculptured surfaces. The designers like smooth transitions, and love the shape behavior close to

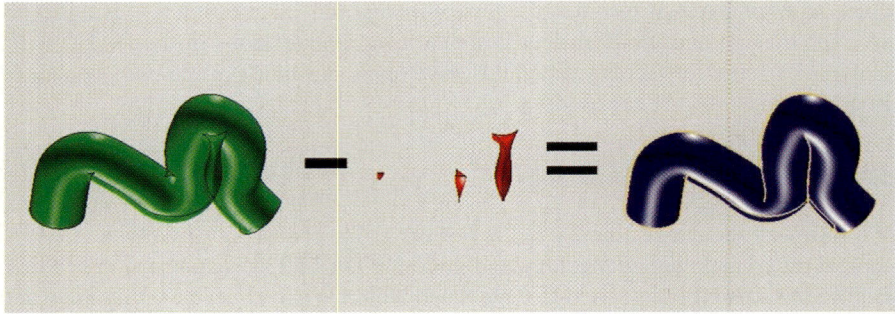

Fig. 1.3. Example from the partner think3 of self-intersection detection and repair integrated into thinkdesign.

surface singularities. However, such shapes often challenge the CAD-systems mathematical basis, especially with respect to surfaces intersecting in a singular or near singular way and surface self-intersections.

Only rudimentary self-intersection software existed in CAD-systems before GAIA II, e.g., rough test to determine that a surface did not contain any self-intersection. However, no code existed for general self-intersections and finding their topology and geometry.

1.9.2 After GAIA II: Possible to find the topology and geometric description of self-intersections

The GAIA II project prototypes have demonstrated that it is possible to handle singular and near singular intersections, as well as determine the topology of self-intersections in surfaces, see Figure 1.3. However, the prototypes also demonstrate that we are far from the ultimate perfect solution. For the GAIA II results to get a direct impact on the worldwide CAD-industry, the vendors have to feel that they loose market shares if the technology of GAIA is not integrated to their product. For the GAIA II results to have a significant industrial impact CAD-vendors have to introduce self-intersection algorithms and improved intersection algorithms into their systems. A more indirect impact on the market can be done by suppling plug-ins to major CAD-systems.

The cooperation between CAGD and Algebraic geometry has opened a new research domain in between CAGD and Algebraic geometry, and shown that many challenges within computer based geometry processing remains.

1.9.3 Future outlook: Acceleration of self-intersection algorithms by graphics cards and multi-core algorithms

Moore's law (from 1965) is a rule of thumb in the computer industry about the growth of computing power over time. Attributed to Gordon E. Moore the co-founder of

Intel, it states that the growth of computing power follows an empirical exponential law. Moore originally proposed a 12 month doubling and, later, a 24 month period.

Until recently the evolution of the frequency of the CPU has had a close relation to a doubling every 12, 18 or 24 month. However, in the last years multi-core CPUs have been introduced. As long as the growth in computational power was related to the CPU-frequency, old sequential program codes could easily profit from the growth in computational power. However, with multi-core CPUs the code has to be prepared for multi-core CPUs to benefit from the performance. Consequently, the era when old sequential program codes automatically benefit from Moore's law is coming to an end. In the coming years reimplementation of algorithms will be necessary to benefit significantly from Moores law.

The GAIA II results have shown significantly improvements in CAD-functionality, but we have also experienced that the 2005 level single-core CPUs are too slow for efficient industrial use of the results. However, with the ongoing activity within SINTEF on GPU-acceleration of intersection algorithms and the use of multi-core CPUs will make accessible sufficient low cost computational resources for industrial use of the GAIA II results. SINTEF has already started on this work [5] as stated above, and has addressed IPR-protection by patenting.

The ideas of GAIA II should be combined with GPU-acceleration and multi-core CPUs. There are indications that visualization and simulation will be central in FP7. If this is the case GAIA II and the SINTEF GPU-activity can be viewed as preproject for proposals within FP7.

1.9.4 Future outlook: More use of algebraic representations in CAD

Although we have not found as much results in traditional real algebraic geometry as expected to be used within CAD, the work on approximate implicitization and approximate parameterization has opened a bridge between parametric and algebraic representation that earlier did not exist. We also expect that more efficient visualization techniques will be available for algebraic surfaces in the years coming. When this is in place we expect a much wider use of algebraic geometry both in CAD and in applications within petroleum and health.

1.9.5 Use of the GAIA II results by other researchers in the area

With the broad range of papers published by GAIA II project partners, most of the research done within GAIA II is already available to other researchers in the area. The reference list following contains papers related to the GAIA II project published by the partners form the start of the GAIA assessment project until the publication of this book.

Much of the most important software of GAIA II is already available or will be available for download on the Internet as Open Source (GNU GPL License):

- AXEL library is available at http://www-sop.inria.fr/galaad/.
- Approximate implicitization is available at http://www.sintef.no/math/.

- The combined approximate implicit and recursive intersection toolkit is planned to be available second half of 2006 from http://www.sintef.no/math/.

Thus most of the results interesting to researchers will be available, and can be a starting point for further research. As also software tools are/will be available researchers can start directly from the GAIA II algorithms implemented and avoid re-implementing the algorithms of GAIA II before their research starts.

References

1. M. Aigner and B. Jüttler. Robust computation of foot points on implicitly defined curves. In M. Dæhlen, K. Mørken, and L. Schumaker, editors, *Mathematical Methods for Curves and Surfaces: Tromsø 2004*, pages 1–10. Nashboro Press, Brentwood, 2005.
2. M. Aigner, B. Jüttler, and Myung-Soo Kim. Analyzing and enhancing the robustness of implicit representations. In *Geometric Modelling and Processing*, pages 131–140. IEEE Press, 2004.
3. M. Aigner, I. Szilagyi, J. Schicho, and B. Jüttler. Implicitization and distance bounds. In M. Elkadi, B. Mourrain, and R. Piene, editors, *Algebraic Geometry and Geometric Modeling*, Mathematics and Visualization. Springer, 2006.
4. J. G. Alcázar and J. R. Sendra. Computation of the topology of real algebraic space curves. *J. Symbolic Comput.*, 39(6):719–744, 2005.
5. S. Briseid, T. Dokken, T. R. Hagen, and J. O. Nygaard. Spline surface intersections optimized for gpus. In V. N. Alexandrov, G. D. van Albada, P. M. A. Sloot, and J. Dongarra, editors, *Computational Science ICCS 2006: 6th International Conference, Reading, UK, May 28-31, 2006, Proceedings, Part IV*, volume 3994 of *Lecture Notes in Computer Science*, pages 204 – 211, Berlin / Heidelberg, 2006. ICCS, Springer.
6. L. Busé and M. Chardin. Implicitizing rational hypersurfaces using approximation complexes. *J. Symbolic Comput.*, 40(4-5):1150–1168, 2005.
7. L. Busé and C. D'Andrea. Inversion of parameterized hypersurfaces by means of subresultants. In *ISSAC 2004*, pages 65–71. ACM, New York, 2004.
8. L. Busé, M. Elkadi, and B. Mourrain. Using projection operators in computer aided geometric design. In *Topics in algebraic geometry and geometric modeling*, volume 334 of *Contemp. Math.*, pages 321–342. Amer. Math. Soc., Providence, RI, 2003.
9. L. Busé and A. Galligo. Using semi-implicit representation of algebraic surfaces. In *Proceedings of IEEE International Conference on Shape Modeling and Applications 2004 (SMI'04)*, pages 342–345, Los Alamitos, CA, USA, 2004. IEEE Computer Society.
10. F. Carreras and L. Gonzalez-Vega. A bisection scheme for intersecting implicit curves. In *Proceedings of the Encounters of Computer Algebra and its Applications EACA-2004, 77-82, Universidad de Cantabria, 2005*, To appear.
11. P. Chalmovianský and B. Jüttler. Filling holes in point clouds. In M. Wilson and R.R. Martin, editors, *The Mathematics of Surfaces X*, volume 2768 of *Lecture Notes in Computer Science*, pages 196–212, Berlin, 2003. Springer.
12. P. Chalmovianský and B. Jüttler. Approximate parameterization by planar rational curves. In *20th Spring Conference on Computer Graphics*, pages 27–35. Comenius University / ACM Siggraph, 2004.
13. P. Chalmovianský and B. Jüttler. Fairness criteria for algebraic curves. *Computing*, 72:41–51, 2004.

14. S. Chau, M. Oberneder, A. Galligo, and B. Jüttler. Intersecting biquadratic Bézier surface patches. In this volume.

15. R. M. Corless, L. Gonzalez-Vega, I: Necula, and A. Shakoori. Topology determination of implicitly defined real algebraic plane curves. *An. Univ. Timişoara Ser. Mat.-Inform.*, 41(Special issue):83–96, 2003.

16. T. Dokken. *Aspect of Intersection algorithms and Approximation, Thesis for the doctor philosophias degree.* PhD thesis, University of Oslo, 1997.

17. T. Dokken. Approximate implicitization. In *Mathematical methods for curves and surfaces (Oslo, 2000)*, Innov. Appl. Math., pages 81–102. Vanderbilt Univ. Press, Nashville, TN, 2001.

18. T. Dokken. Controlling the shape of the error in cubic ellipse approximation. In *Curve and surface design (Saint-Malo, 2002)*, Mod. Methods Math., pages 113–122. Nashboro Press, Brentwood, TN, 2003.

19. T. Dokken and B. Jüttler, editors. *Computational Methods for Algebraic Spline Surfaces.* Springer, Heidelberg, 2005.

20. T. Dokken and V. Skytt. Intersection algorithms and cagd. In *Applied Mathematics at SINTEF*. Springer, To appear.

21. T. Dokken and J. B. Thomassen. Overview of approximate implicitization. In *Topics in algebraic geometry and geometric modeling*, volume 334 of *Contemp. Math.*, pages 169–184. Amer. Math. Soc., Providence, RI, 2003.

22. T. Dokken and J. B. Thomassen. Weak approximate implicitization. In *Proceedings of IEEE International Conference on Shape Modeling and Applications 2006 (SMI'06)*, pages 204–214, Los Alamitos, CA, USA, 2006. IEEE Computer Society.

23. M. Elkadi, A. Galligo, and T.H. Lê. Parametrized surfaces in $\mathbb{P}^3$ of bidegree $(1, 2)$. In *ISSAC 2004*, pages 141–148. ACM, New York, 2004.

24. M. Elkadi and B. Mourrain. Residue and implicitization problem for rational surfaces. *Appl. Algebra Engrg. Comm. Comput.*, 14(5):361–379, 2004.

25. F. Etayo, L. Gonzalez-Vega, and N. del Rio. A complete solution for the ellipses intersection problem. *Comput. Aided Geom. Design*, 23(4):324–350, 2006.

26. F. Etayo, L. González-Vega, and C. Tănăsescu. Computing the intersection curve of two surfaces: The tangential case. *An. Univ. Timişoara Ser. Mat.-Inform.*, 41(Special issue):111–121, 2003.

27. M. Fioravanti and L. Gonzalez-Vega. On the geometric extraneous components appearing when using implicitization. In *Mathematical methods for curves and surfaces: Tromsø 2004*, Mod. Methods Math., pages 157–168. Nashboro Press, Brentwood, TN, 2005.

28. M. Fioravanti, L. Gonzalez-Vega, and I. Necula. Computing the intersection of two ruled surfaces by using a new algebraic approach. *J. Symbolic Comput.*

29. M. Fioravanti, L. Gonzalez-Vega, and I. Necula. On the intersection with revolution and canal surfaces. In M. Elkadi, B. Mourrain, and R. Piene, editors, *Algebraic Geometry and Geometric Modeling*, Mathematics and Visualization. Springer, 2006.

30. G. Gatellier, A. Labrouzy, B. Mourrain, and J. P. Técourt. Computing the topology of three-dimensional algebraic curves. In *Computational methods for algebraic spline surfaces*, pages 27–43. Springer, Berlin, 2005.

31. L. González-Vega and I. Necula. Efficient topology determination of implicitly defined algebraic plane curves. *Comput. Aided Geom. Design*, 19(9):719–743, 2002.

32. L. González-Vega, I. Necula, S. Pérez-Díaz, J. Sendra, and J. R. Sendra. Algebraic methods in computer aided geometric design: theoretical and practical applications. In F. Chen and D. Wang, editors, *Geometric computation*, volume 11 of *Lecture Notes Series on Computing*, pages 1–33. World Sci. Publishing, River Edge, NJ, 2004.

33. L. González-Vega, I. Necula, and J. Puig-Pey. Manipulating 3d implicit surfaces by using differential equation solving and algebraic techniques. In F. Chen and D. Wang, editors, *Geometric computation*, volume 11 of *Lecture Notes Series on Computing*. World Scientific Publishing, River Edge, NJ, 2004.

34. P. H. Johansen. The geometry of the tangent developable. In *Computational methods for algebraic spline surfaces*, pages 95–106. Springer, Berlin, 2005.

35. P. H. Johansen, M. Løberg, and R. Piene. Monoid hypersurfaces. In this volume.

36. B. Jüttler, P. Chalmoviansky, M. Shalaby, and E. Wurm. Approximate algebraic methods for curves and surfaces and their applications. In *21st Spring Conference on Computer Graphics*. Comenius University / ACM Siggraph, 2005.

37. B. Jüttler, J. Schicho, and M. Shalaby. Spline implicitization of planar curves. In *Curve and surface design (Saint-Malo, 2002)*, Mod. Methods Math., pages 225–234. Nashboro Press, Brentwood, TN, 2003.

38. B. Jüttler, J. Schicho, and M. Shalaby. C^1 spline implicitization of planar curves. In F. Winkler, editor, *Automated deduction in geometry*, Lecture Notes in Artificial Intelligence, pages 161–177, Heidelberg, 2004. Springer.

39. B. Jüttler and W. Wang. The shape of spherical quartics. *Computer Aided Geometric Design*, 20:621–636, 2003.

40. B. Jüttler and E. Wurm. Approximate implicitization via curve fitting. In L. Kobbelt, P. Schröder, and H. Hoppe, editors, *Symposium on Geometry Processing*, pages 240–247, New York, 2003. Eurographics / ACM Siggraph.

41. T. H. Le and A. Galligo. General classification of (1,2) parametric surfaces in p^3, 2006. In this volume.

42. B. Mourrain. Bezoutian and quotient ring structure. *J. Symbolic Comput.*, 39(3-4):397–415, 2005.

43. S. Pérez-Díaz, J. Sendra, and J. R. Sendra. Parametrization of approximate algebraic curves by lines. *Theoret. Comput. Sci.*, 315(2-3):627–650, 2004.

44. S. Pérez-Díaz, J. Sendra, and J. R. Sendra. Distance properties of ϵ-points on algebraic curves. In *Computational methods for algebraic spline surfaces*, pages 45–61. Springer, Berlin, 2005.

45. S. Pérez-Díaz, J. Sendra, and J. R. Sendra. Parametrization of approximate algebraic surfaces by lines. *Comput. Aided Geom. Design*, 22(2):147–181, 2005.

46. S. Pérez-Díaz and J. R. Sendra. Computing all parametric solutions for blending parametric surfaces. *J. Symbolic Comput.*, 36(6):925–964, 2003.

47. S. Pérez-Díaz and J. R. Sendra. Computation of the degree of rational surface parametrizations. *J. Pure Appl. Algebra*, 193(1-3):99–121, 2004.

48. R. Piene. Singularities of some projective rational surfaces. In *Computational methods for algebraic spline surfaces*, pages 171–182. Springer, Berlin, 2005.

49. J. R. Sendra. Rational curves and surfaces: algorithms and some applications. In F. Chen and D. Wang, editors, *Geometric computation*, volume 11 of *Lecture Notes Series on Computing*, pages 65–125. World Sci. Publishing, River Edge, NJ, 2004.

50. M. Shalaby and B. Jüttler. Approximate implicitization of space curves and of surfaces of revolution, 2006. In this volume.

51. M. Shalaby, B. Jüttler, and J. Schicho. Approximate implicitization of planar curves by piecewise rational approximation of the distance function. *Appl. Algebra Eng. Comp.*, to appear.

52. M. Shalaby, J. Thomassen, E. Wurm, T. Dokken, and B. Jüttler. Piecewise approximate implicitization: Experiments using industrial data. In M. Elkadi, B. Mourrain, and R. Piene, editors, *Algebraic Geometry and Geometric Modeling*, Mathematics and Visualization. Springer, 2006.

53. Z. Šír. Approximate parametrisation of confidence sets. In *Computational methods for algebraic spline surfaces*, pages 1–10. Springer, Berlin, 2005.
54. V. Skytt. Challenges in surface-surface intersections. In *Computational methods for algebraic spline surfaces*, pages 11–26. Springer, Berlin, 2005.
55. V. Skytt. A recursive approach to surface-surface intersections. In *Mathematical methods for curves and surfaces: Tromsø 2004*, Mod. Methods Math., pages 327–338. Nashboro Press, Brentwood, TN, 2005.
56. J. B. Thomassen. Self-intersection problems and approximate implicitization. In *Computational methods for algebraic spline surfaces*, pages 155–170. Springer, Berlin, 2005.
57. J. B. Thomassen, P. H. Johansen, and T. Dokken. Closest points, moving surfaces, and algebraic geometry. In *Mathematical methods for curves and surfaces: Tromsø 2004*, Mod. Methods Math., pages 351–362. Nashboro Press, Brentwood, TN, 2005.
58. E. Wings and B. Jüttler. Generating tool paths on surfaces for a numerically controlled calotte cutting system. *Computer-Aided Design*, 36:325–331, 2004.
59. E. Wurm, B. Jüttler, and M.-S. Kim. Approximate rational parameterization of implicitly defined surfaces. In R. Martin, H. Bez, and M. Sabin, editors, *Mathematics of Surfaces XI*, volume 3604 of *LNCS*, pages 434–447. Springer, 2005.
60. E. Wurm, J.B. Thomassen, B. Jüttler, and T. Dokken. Comparative benchmarking of methods for approximate parameterization. In M. Neamtu and M. Lucian, editors, *Geometric Modeling and Computing: Seattle 2003*, pages 537–548. Nashboro Press, Brentwood, 2004.

Part II

Some special algebraic surfaces

The second part of this book contains chapters which describe results concerning special algebraic surfaces. Most surfaces used in geometric modeling are algebraic surfaces of low degree, and their geometric nature, in particular their singularities, can be analyzed using tools from real algebraic geometry. Here we collect several results in this direction, which are organized in five chapters.

Aries, Briand and Bruchou analyze some covariants related to Steiner surfaces, which are the generic case of a quadratically parameterizable quartic surface, frequently used in geometric modeling. More precisely, they exhibit a collection of covariants associated to projective quadratic parameterizations of surfaces with respect to the actions of linear reparameterizations and linear transformations of the target space. Along with the covariants, the authors provide simple geometric interpretations. The results are then used to generate explicit equations and inequalities defining the orbits of projective quadratic parameterizations of quartic surfaces.

The next chapter, authored by Breske, Labs and van Straten, is devoted to real line arrangements and surfaces with many real nodes. It is shown that Chmutov's construction for surfaces with many singularities can be modified so as to give surfaces with only real singularities. The results show that all known lower bounds for the number of nodes can be attained with only real singularities. The paper concludes with an application of the theory of real line arrangements which shows that the arrangements used by the authors are asymptotically the best possible ones for the purpose of constructing surfaces with many nodes. This proves a special case of a conjecture of Chmutov.

Johansen, Løberg and Piene study properties of monoid hypersurfaces – irreducible hypersurfaces of degree d with a singular point of multiplicity $d - 1$. Since such surfaces admit a rational parameterization, they are of potential interest in computer aided geometric design. The main results include a description of the possible real forms of the singularities on a monoid surface other than the $(d - 1)$-uple point. The results are applied to the classification of singularities on quartic monoid surfaces, complementing earlier work on the subject.

The chapter by Krasauskas and Zube discusses canal surfaces which are generated as the envelopes of quadratic families of spheres. These surfaces generalize the class of Dupin cyclides, but they are more flexible as blending surfaces between natural quadrics. The authors provide a classification from the point of view of Laguerre geometry and study rational parameterizations of minimal degree, Bézier representations, and implicit equations.

Finally Lê and Galligo present the classification of surfaces of bidegree (1,2) over the fields of complex and real numbers. In particular, the authors study patches of such surfaces, and they show how to detect and describe the loci in the parameter domain – a $[0, 1] \times [0, 1]$ box – that map to selfintersections and singular points on the surface.

2

Some Covariants Related to Steiner Surfaces

Franck Aries[1], Emmanuel Briand[2], and Claude Bruchou[1]

[1] INRA Biométrie, Avignon (France)
    Franck.Aries@avignon.inra.fr Claude.Bruchou@avignon.inra.fr
[2] Universidad de Cantabria, Santander (Spain)
    emmanuel.briand@gmail.com

Summary. A Steiner surface is the generic case of a quadratically parameterizable quartic surface used in geometric modeling. This paper studies quadratic parameterizations of surfaces under the angle of Classical Invariant Theory. Precisely, it exhibits a collection of covariants associated to projective quadratic parameterizations of surfaces, under the actions of linear reparameterization and linear transformations of the target space. Each of these covariants comes with a simple geometric interpretation.

As an application, some of these covariants are used to produce explicit equations and inequalities defining the orbits of projective quadratic parameterizations of quartic surfaces.

2.1 Introduction

This paper deals with quadratically parameterizable quartic surfaces of $\mathbb{R}^3$, that is surfaces of degree 4 admitting a parameterization of the form:

$$
\begin{array}{ccc}
\mathbb{R}^2 & \longrightarrow & \mathbb{R}^3 \\
(x_1, x_2) & \longmapsto & \left(\frac{F_1(x_1,x_2)}{F_0(x_1,x_2)}, \frac{F_2(x_1,x_2)}{F_0(x_1,x_2)}, \frac{F_3(x_1,x_2)}{F_0(x_1,x_2)} \right)
\end{array}
\tag{2.1}
$$

where the F_i are polynomial functions of degree at most 2. For generic F_i's, the parameterized surface obtained is called a *Steiner surface*, see section 2.2 for the precise definition.

Our general motivation for the study of Steiner surfaces is the following. Two of us (Franck Aries and Claude Bruchou) are interested in mathematical modeling of vegetation canopies (see [9] for more details). The detailed description of the architecture of vegetation canopies is critical for the modeling of many agricultural processes: the photosynthesis, the propagation of diseases from one organ to another or the radiative transfer. These processes involve a big amount of computations on geometric objects associated to each plant organ. Each geometric object can be approximated by a set of plane triangles, or more complex patches like bicubic. As underlined in several papers of geometric modeling ([2, 7, 15]), Steiner patches are a possibly good compromise between triangles, which need to be very many for a good accuracy, and the eighteen degree surfaces associated to the bicubic parameterization, which raise problems of complexity. Unfortunately, one may meet singular, or close to singular parameterizations, that make computations unreliable. Thus one needs to know as much as possible about the geometry of the space of quadratic parameterizations.

The study of quadratic parameterizations is eased by considering, instead of the affine setting, the projective setting. This means considering the projective quadratic parameterizations of surfaces, that is the quadratic rational maps from the real projective plane $\mathbb{RP}^2$ to the real projective space $\mathbb{RP}^3$. These maps are those of the form:

$$
\begin{aligned}
\Omega \subset \mathbb{RP}^2 &\longrightarrow \mathbb{RP}^3 \\
(x_0 : x_1 : x_2) &\longmapsto (f_0 : f_1 : f_2 : f_3) \, .
\end{aligned}
\tag{2.2}
$$

where the f_i are quadratic forms in x_0, x_1, x_2 and Ω is a non–empty Zariski open subset of $\mathbb{RP}^2$.

The main topic of the present paper is the Invariant Theory of projective quadratic parameterizations under linear changes of coordinates of $\mathbb{RP}^2$ and $\mathbb{RP}^3$. Precisely, we provide a collection of covariants with simple geometric interpretation.

Let us give a motivating problem: the discrimination between the different kinds of quadratic parameterizations of quartic surfaces. Let us make this precise. Consider a quadratic map as in (2.2). Its image in $\mathbb{RP}^3$ is not, in general, Zariski–closed. Consider its Zariski closure, it is an algebraic surface of degree at most 4. Let $\mathcal{U}$ be the set of those maps for which it is a quartic, *i.e.* it has degree exactly 4. Two elements of $\mathcal{U}$ are considered *equivalent* if one is obtained from the other by a linear reparameterization (linear change of coordinates in the domain $\mathbb{RP}^2$) and a projective transformation of the ambient space (linear change of coordinates in the codomain $\mathbb{RP}^3$). Then, as it is shown in [7] and [8], there are finitely many equivalence classes in $\mathcal{U}$. The problem is to discriminate between these equivalence classes. Algorithmic solutions to this problem have been given in [2] and [7]. Our paper proposes a new solution. It consists simply in providing polynomial equations and inequalities defining the equivalence classes[3]. The equivalence classes are actually orbits under the action of some group. Thus it is natural to look for the equations and inequalities among the objects provided by Classical Invariant Theory: the covariants. Then, the aforementioned problem of discrimination between orbits of parameterizations is solved as an application, by picking in our toolbox of covariants the most adapted ones.

The sequel of the paper is organized as follows: Section 2.2 recalls known facts about the classification of quadratic parameterizations of surfaces; Section 2.3 provides preliminaries on Classical Invariant Theory; Section 2.4 presents some geometrical features of Steiner surfaces, that will be helpful to present our collection of covariants; these covariants are introduced in Section 2.5; the last section, Section

---

[3] Here is an example where the methods of [2] and [7] are not directly applicable: suppose we are given a family of parameterizations, depending on a parameter t. Then, by mere specialization of the general equations and inequalities defining the classes, we are able to determine which values of t give a parameterization in a given equivalence class.

2.6, presents the application of these covariants to the discrimination of classes of parameterizations.

2.2 Orbits of quadratic parameterizations of quartics

A quadratic rational map from $\mathbb{RP}^2$ to $\mathbb{RP}^3$ is determined by a homogeneous quadratic map f from $\mathbb{R}^3$ to $\mathbb{R}^4$, that can be presented as a family of four real ternary quadratic forms:

$$f = (f_0(x_0, x_1, x_2), f_1(x_0, x_1, x_2), f_2(x_0, x_1, x_2), f_3(x_0, x_1, x_2)). \qquad (2.3)$$

Denote with $\mathcal{F}$ the space of all the quadruples of real ternary quadratic forms. Then, more precisely, quadratic rational maps from $\mathbb{RP}^2$ to $\mathbb{RP}^3$ can be identified with the elements of $\mathcal{F}$ considered *modulo* scalar multiplication, *i.e.* the projective space $\mathbb{P}(\mathcal{F})$. For $f \in \mathcal{F}$, we will denote with $[f]$ the corresponding element of $\mathbb{P}(\mathcal{F})$.

Now the group $GL(3, \mathbb{R})$ acts naturally on $\mathbb{R}^3$ (and $\mathbb{RP}^2$), and thus on $\mathcal{F}$ (and $\mathbb{P}(\mathcal{F})$). The action on $\mathcal{F}$ is as follows: for $\theta \in GL(3, \mathbb{R})$,

$$\theta(f) = f \circ \theta^{-1}. \qquad (2.4)$$

The induced action on $\mathbb{P}(\mathcal{F})$ corresponds to linear reparameterizations. There is also a natural action of the group $GL(4, \mathbb{R})$ on $\mathbb{R}^4$ (and $\mathbb{RP}^3$), and thus on $\mathcal{F}$ (and $\mathbb{P}(\mathcal{F})$): for $\rho \in GL(4, \mathbb{R})$,

$$\rho(f) = \rho \circ f. \qquad (2.5)$$

We have thus an action of $GL(3, \mathbb{R}) \times GL(4, \mathbb{R})$ on $\mathcal{F}$ (and $\mathbb{P}(\mathcal{F})$). In the sequel, we will denote this group with G.

In $\mathbb{P}(\mathcal{F})$, the subset $\mathcal{U}$ of those projective parameterizations with the property that the Zariski closure of their image[4] is a surface of degree 4 exactly, is invariant under G. It is also a Zariski dense open set. As said in the introduction, the decomposition of $\mathcal{U}$ into orbits is known[5]; see [2, 7] and [8]. There are only six orbits. Table 2.1 provides the list of the orbits, with a representative for each.

Let us say a word about the connection between this problem and the analogous problem in the complex setting. Denote with $\mathcal{F}_{\mathbb{C}}$ the complexification of $\mathcal{F}$: that is the space of families of four complex quadratic forms. Then $\mathbb{P}(\mathcal{F}_{\mathbb{C}})$ represents the space of quadratic rational maps from the complex projective plane, $\mathbb{CP}^2$ to the complex projective three–dimensional space, $\mathbb{CP}^3$. Let $\mathcal{U}_{\mathbb{C}}$ be the subset of those parameterizations whose image is a quartic surface. Then $\mathcal{U}$ is the trace of $\mathcal{U}_{\mathbb{C}}$ on $\mathbb{P}(\mathcal{F})$. This means that $\mathcal{U} = \mathcal{U}_{\mathbb{C}} \cap \mathbb{P}(\mathcal{F})$.

Let $G_{\mathbb{C}} = GL(3, \mathbb{C}) \times GL(4, \mathbb{C})$. This group acts naturally on $\mathcal{F}_{\mathbb{C}}$ and $\mathbb{P}(\mathcal{F}_{\mathbb{C}})$, and also on $\mathcal{U}_{\mathbb{C}}$. The classification of the orbits of $\mathbb{P}(\mathcal{F})$ under G is obtained by refining the classification of $\mathbb{P}(\mathcal{F}_{\mathbb{C}})$ into orbits under $G_{\mathbb{C}}$ (see [1] for a modern reference about

---

[4] We consider the set–theoretical image, and rule out the cases when the Zariski closure of the image is a double quadric (case 7 in Proposition 5 of [2]) or a plane counted four times.

[5] The determination of the orbits outside $\mathcal{U}$ is a different problem. See the references in [7].

Orbit	Representative
Ii	$\left(2\,x_1x_2 : 2\,x_0x_2 : 2\,x_0x_1 : x_0{}^2 + x_1{}^2 + x_2^2\right)$
Iii	$\left(2\,x_1x_2 : 2\,x_0x_2 : 2\,x_0x_1 : x_0{}^2 - x_1^2 + x_2^2\right)$
Iiii	$\left(x_0{}^2 + x_1^2 : x_1^2 + x_2^2 : x_0x_2 : x_1x_2\right)$
IIi	$\left(x_0{}^2 - x_1^2 : x_0x_1 : x_1x_2 : x_2^2\right)$
IIii	$\left(x_0x_2 - x_1x_2 : x_0{}^2 : x_1^2 : x_2^2\right)$
III	$\left(x_0{}^2 : x_0x_2 - x_1^2 : x_1x_2 : x_2^2\right)$

Table 2.1. Orbits of quadratic parameterizations of quartic surfaces.

this classification in the complex setting). Precisely: if $\mathcal{O}$ is an orbit in $\mathbb{P}(\mathcal{F}_{\mathbb{C}})$ under $G_{\mathbb{C}}$, then its trace (intersection with $\mathbb{P}(\mathcal{F})$) is a union of orbits under G. For instance, $\mathcal{U}_{\mathbb{C}}$ decomposes in three orbits: $I_{\mathbb{C}}$, $II_{\mathbb{C}}$ and $III_{\mathbb{C}}$, and their respective traces on $\mathcal{U}$ are Ii $\cup$ Iii $\cup$ Iiii, IIi $\cup$ IIii, and III.

It happens that there is one dense orbit in $\mathbb{P}(\mathcal{F}_{\mathbb{C}})$: that is Orbit $I_{\mathbb{C}}$. Then a *complex Steiner surface* is just the image in $\mathbb{CP}^3$ of a parameterization in this orbit[6]. It is always a Zariski closed quartic surface. By extension, the name "Steiner surface" is sometimes used for the set of its real points[7]; that is a real quartic surface, Zariski closure of the image of a parameterization in Orbit Ii, Iii or Iiii.

2.3 Preliminaries on classical invariant theory

The objects we will introduce in Section 2.5 are *polynomial covariants* for the action of G on $\mathcal{F}$. We wish now to recall the general definition (we point out [11] and [12] as modern references for Classical Invariant Theory).

Let $\mathcal{G}$ be a group (we will apply what follows for $\mathcal{G} = G$), and let W be some finite-dimensional $\mathcal{G}$–module, that is: a vector space on which $\mathcal{G}$ acts linearly (we will have $W = \mathcal{F}$). Let V be another finite-dimensional $\mathcal{G}$–module. A *polynomial covariant*[8] *of W of type V* is a polynomial map C from W to V, equivariant with respect to $\mathcal{G}$. This means that:

$$C(g(w)) = g(C(w)) \quad \forall w \in W, \quad \forall g \in \mathcal{G}. \tag{2.6}$$

This includes the (relative) invariants, which are the polynomial functions I on W such that for all $g \in \mathcal{G}$, there exists some scalar $c(g)$ such that:

$$I(g(w)) = c(g) \cdot I(w) \quad \forall w \in W. \tag{2.7}$$

---

[6] One could, following some sources in the literature, refer to surfaces in Orbits $II_{\mathbb{C}}$ and $III_{\mathbb{C}}$ as "degenerate" Steiner surfaces, but we will use the term Steiner surface only for the non–degenerate case, *i.e.* only for the elements of Orbit $I_{\mathbb{C}}$.

[7] Nevertheless *Steiner's Roman surface* properly said corresponds to the Zariski closure of the image of a parameterization in Orbit Ii; see [7].

[8] This is the modern meaning for *covariant*, which includes the classical notions of covariants, contravariants and mixed concomitants.

For $\mathcal{G} = G$ acting on $W = \mathcal{F}$, a polynomial covariant for the action of G on $\mathcal{F}$ is a polynomial map from $\mathcal{F}$ to some G–module such that

$$C(\rho \circ f \circ \theta^{-1}) = (\rho, \theta)\,(C(f)) \qquad (2.8)$$

for all $\theta \in GL(3, \mathbb{R})$ and all $\rho \in GL(4, \mathbb{R})$.

Note that the zero set of any covariant is a $\mathcal{G}$–invariant set, that is a union of orbits.

We finish this section with some remarks. The covariants for $\mathcal{F}$ under G are essentially the same as those of $\mathcal{F}_{\mathbb{C}}$ under $G_{\mathbb{C}}$: the former are obtained by complexification of the latter[9]. From a classical theorem of Invariant Theory (see [12]), we know that the homogeneous covariants separate the orbits of $P(\mathcal{F}_{\mathbb{C}})$ under $G_{\mathbb{C}}$: this means that for any two orbits $\mathcal{O}_1$ and $\mathcal{O}_2$, there exists some homogeneous covariant vanishing on $\mathcal{O}_1$ and not on $\mathcal{O}_2$, or *vice–versa*. On the contrary, there is no guarantee in advance that we can separate the orbits of $\mathbb{P}(\mathcal{F})$ under G using equations and inequalities involving only the covariants. We will be able to do it in Section 2.6 by using some derived objects.

2.4 Some elements of the geometry of the Steiner surface

To each of the covariants we will introduce is attached a simple geometric object associated to the quadratic parameterizations of the complex Steiner surface. This is, actually, what will guide us in the construction of the covariants.

We now introduce the main features of the Steiner surface (they can be found in [14], parag. 554a). For $f \in \mathcal{F}$, denote with $S(f)$ the associated complex Steiner surface, that is the image of $\mathbb{CP}^2$ under $[f]$. Then:

- It is a quartic (its implicit equation has degree 4).
- Its singular locus is the union of three lines, that are double lines. They are concurrent: their intersection is the unique triple point of the Steiner surface.
- The intersection of $S(f)$ with a tangent plane is a quartic curve that either decomposes as the union of two conics intersecting at four points, or as a double conic. The latter situation happens only for four tangent planes, that Salmon calls *tropes*. In the former situation, one of the four intersection points is the point of tangency; the three remaining points are the intersections of the plane with each of three double lines.
- Each trope is tangent to the Steiner surface along a conic, called a *torsal conic*[10]. There are thus four torsal conics.
- There is a unique quadric going through the four torsal conics. Let us call it *the Associated Quadric*.
- The dual (or "reciprocal") surface to $S(f)$ (the surface of $(\mathbb{CP}^3)^*$ that is the Zariski closure of the set of all tangent planes to $S(f)$) is a cubic surface, known as the *Cayley Cubic Surface* (see [14]).

---

[9] For such issues of field of definition, see [11].
[10] This is called a *parabolic* conic in [7].

Also of interest are some facts connected to the quadratic parameterization $[f]$ (rather than to the Steiner surface $\mathcal{S}(f)$ itself):

- It is defined on the whole $\mathbb{CP}^2$.
- The direct image of each line of $\mathbb{CP}^2$ is a conic on $\mathcal{S}(f)$.
- The preimage of each conic drawn on $\mathcal{S}(f)$ is a straight line of $\mathbb{CP}^2$. As a consequence, the preimage of any tangent plane is a pair of lines. The lines are distinct, unless the plane is a trope.
- The four lines obtained as preimages of the four tropes (equivalently: of the torsal conics; yet equivalently: of the Associated Quadric) form a non–degenerate quadrilateral.
- The preimage of each of the singular lines of $\mathcal{S}(f)$ is a straight line of $\mathbb{CP}^2$. The 3 lines obtained this way are non concurrent: they form a (non–degenerate) triangle, that we call the *Exceptional Triangle*.
- The preimage of the triple point is the union of the vertices of the Exceptional Triangle.
- The parameterization is faithful (*i.e.* generically injective). Precisely, it is injective on the complement of the Exceptional Triangle in $\mathbb{CP}^2$.

2.5 A collection of covariants

2.5.1 Preliminaries

This section presents the new contribution of the paper: a collection of homogeneous covariants for the action of G on $\mathcal{F}$, with a simple geometric interpretation for each of them.

Let us start with some notations. Denote the canonical basis of $\mathbb{C}^3$ with λ_0, λ_1, λ_2 and its dual basis with x_0, x_1, x_2. Denote also the canonical basis of $\mathbb{C}^4$ with α_0, α_1, α_2, α_3 and its dual basis with y_0, y_1, y_2, y_3. Given two complex vector spaces W and V, denote with $\mathsf{Pol}_n(W, V)$ the space of homogeneous polynomial maps from W to V of degree n. Denote also $\mathsf{Pol}_n(W)$ the space of polynomial homogeneous functions of degree n over W. Otherwise stated,

$$\mathsf{Pol}_n(W) = \mathsf{Pol}_n(W, \mathbb{C}). \tag{2.9}$$

For $f = (f_0, f_1, f_2, f_3) \in \mathcal{F}$, denote the coefficients of f_i with a_{ij} and b_{ij}, as follows:

$$f_i = a_{i0}\, x_0^2 + a_{i1}\, x_1^2 + a_{i2}\, x_2^2 + 2\, b_{i0}\, x_1\, x_2 + 2\, b_{i1}\, x_0\, x_2 + 2\, b_{i2}\, x_0\, x_1. \tag{2.10}$$

Each of the homogeneous covariants we will present, considered up to a scalar, represents some geometric object associated to the parameterization $[f]$, according to its type (its space of values[11]). Note that the definition of this geometric object will be valid only in the case when $[f]$ parameterizes a Steiner surface.

We will meet covariants of the following types:

- Type $\mathsf{Pol}_n(\mathbb{C}^4)$: such a covariant C associates to $[f]$ a surface in $\mathbb{CP}^3$ (the zero locus of $C(f)$).
- Type $\mathsf{Pol}_n(\mathbb{C}^3)$: such a covariant associates to $[f]$ a curve in $\mathbb{CP}^2$.
- Type $\mathsf{Pol}_n((\mathbb{C}^4)^*)$: such a covariant associates to $[f]$ a surface in $(\mathbb{CP}^3)^*$. If this surface is decomposable, that is a union of hyperplanes of $(\mathbb{CP}^3)^*$, then it also represents a finite collection of points in $\mathbb{CP}^3$ (the points corresponding to the hyperplanes by duality).
- Type $\mathsf{Pol}_n((\mathbb{C}^3)^*)$: such a covariant associates to $[f]$ a curve in $(\mathbb{CP}^2)^*$. If this curve is decomposable, then it also represents a finite collection of points in $\mathbb{CP}^2$.
- Type some space of functions $\mathsf{Pol}_n(W, V)$ between spaces W, V among $\mathbb{C}^3$, $\mathbb{C}^4$ and their duals. Then the covariant associates to $[f]$ some family of curves or surfaces in $\mathbb{P}(V)^*$ parameterized by $\mathbb{P}(W)$.
- Type $\mathbb{C}$: such a homogeneous covariant is just an invariant for the group $SL(3, \mathbb{C}) \times SL(4, \mathbb{C})$. We will see that there is essentially only one invariant.

The geometric objects attached to some of the covariants we will present will be clear from their construction; for the rest, they can be found merely by evaluating the covariant on the representative of Orbit $I_{\mathbb{C}}$:

$$\left(2\,x_1 x_2 : 2\,x_0 x_2 : 2\,x_0 x_1 : x_0{}^2 + x_1{}^2 + x_2^2\right). \tag{2.11}$$

Table 2.2 recapitulates the list of covariants that will be now presented individually. The reader will find Maple procedures implementing the formulas that follow on the web page:

http://emmanuel.jean.briand.free.fr/publications/steiner/

2.5.2 Derivation of the covariants

Here we suppose that $[f]$ is in $I_{\mathbb{C}}$, that is its image $\mathcal{S}(f)$ in $\mathbb{CP}^3$ is a complex Steiner surface.

For each covariant we indicate its type, and its degree with respect to the coefficients of the f_i's.

Tangent plane at the image of a point.

Given a generic point $[x]$ in the parameter space $\mathbb{CP}^2$, we can consider the tangent plane to the Steiner surface $\mathcal{S}(f)$ at its image by $[f]$. It has equation $\Phi_1(f)(x) = 0$, where

---

[11] Strictly speaking, the type should mention also the action of G on this space. In all the cases we will meet, this action is a canonical action of G on the space, or its product by some powers of the determinants of $\theta \in GL(3, \mathbb{R})$ and $\rho \in GL(4, \mathbb{R})$. These powers are easily determined from the degree of the covariant.

Symbol	Name	Degree	Type
	Invariants		$\mathbb{C}$
Δ	Discriminant	24	$\mathbb{C}$
	Families of objects		
Φ_1	Tangent plane at the image of a point	3	$\mathsf{Pol}_3(\mathbb{C}^3, (\mathbb{C}^4)^*)$
Φ_2	Linear plane spanned by the image of a line	3	$\mathsf{Pol}_3((\mathbb{C}^3)^*, (\mathbb{C}^4)^*)$
Φ_3	Correspondence line–line	4	$\mathsf{Pol}_2((\mathbb{C}^3)^*, (\mathbb{C}^3)^*)$
Φ_6	Preimage of a point on $\mathcal{S}(f)$	10	$\mathsf{Pol}_2(\mathbb{C}^4, \mathbb{C}^3)$
	Associated surfaces in $\mathbb{CP}^3$		$\mathsf{Pol}_n(\mathbb{C}^4)$
Φ_4	Implicit Equation	12	$n = 4$
Φ_5	Associated Quadric	6	$n = 2$
Φ_9	Union of the Tropes	12	$n = 4$
Φ_{10}	Trihedron defined by the Double Lines	21	$n = 3$
Φ_{12}	Polar Plane Π of the Associated Quadric and the Triple Point	15	$n = 1$
	Associated surfaces in $(\mathbb{CP}^3)^*$		$\mathsf{Pol}_n((\mathbb{C}^4)^*)$
Φ_7	Dual surface	3	$n = 3$
Φ_8	Triple Point	9	$n = 1$
	Associated curves in $\mathbb{CP}^2$		$\mathsf{Pol}_n(\mathbb{C}^3)$
Φ_{11}	Exceptional Triangle	12	$n = 3$
Φ_{13}	Conic preimage of Π	16	$n = 2$
Φ_{15}	Quadrilateral preimage of the torsal conics	8	$n = 4$
	Associated surfaces of $(\mathbb{CP}^2)^*$		$\mathsf{Pol}_n((\mathbb{C}^3)^*)$
Φ_{14}	Dual conic to the preimage of Π	8	$n = 2$

Table 2.2. List of the covariants presented in the paper.

$$\Phi_1 = \frac{1}{8} \begin{vmatrix} \partial_0 f_0 & \partial_1 f_0 & \partial_2 f_0 & y_0 \\ \partial_0 f_1 & \partial_1 f_1 & \partial_2 f_1 & y_1 \\ \partial_0 f_2 & \partial_1 f_2 & \partial_2 f_2 & y_2 \\ \partial_0 f_3 & \partial_1 f_3 & \partial_2 f_3 & y_3 \end{vmatrix}. \tag{2.12}$$

Here ∂_i stands for $\frac{\partial}{dx_i}$.

This covariant Φ_1 has degree 3 and type $\mathsf{Pol}_3(\mathbb{C}^3, (\mathbb{C}^4)^*)$. The geometric object associated to $\Phi_1(f)$ is a parameterization of the dual surface to $\mathcal{S}(f)$.

Plane spanned by the image of a line.

Consider a generic line L in $\mathbb{CP}^2$, given by an equation

$$\lambda(x) = \lambda_0 x_0 + \lambda_1 x_1 + \lambda_2 x_2 = 0. \tag{2.13}$$

Its image under f is a conic in $\mathbb{CP}^3$, spanning a plane, that is an element of $(\mathbb{CP}^3)^*$. This plane is always a tangent plane to $\mathcal{S}(f)$. It admits $\Phi_2(f)(\lambda) = 0$ as an equation, with

$$\Phi_2 = \frac{1}{2} \begin{vmatrix} a_{00} & a_{01} & a_{02} & 2\,b_{00} & 2\,b_{01} & 2\,b_{02} & y_0 \\ a_{10} & a_{11} & a_{12} & 2\,b_{10} & 2\,b_{11} & 2\,b_{12} & y_1 \\ a_{20} & a_{21} & a_{22} & 2\,b_{20} & 2\,b_{21} & 2\,b_{22} & y_2 \\ a_{30} & a_{31} & a_{32} & 2\,b_{30} & 2\,b_{31} & 2\,b_{32} & y_3 \\ \lambda_0 & 0 & 0 & 0 & \lambda_2 & \lambda_1 & 0 \\ 0 & \lambda_1 & 0 & \lambda_2 & 0 & \lambda_0 & 0 \\ 0 & 0 & \lambda_2 & \lambda_1 & \lambda_0 & 0 & 0 \end{vmatrix}. \tag{2.14}$$

Note that the lines of the matrix in the determinant correspond to the equations:

$$\begin{aligned} f_i(x) &= y_i, \quad i = 0,1,2,3, \\ x_j \lambda(x) &= 0, \quad j = 0,1,2, \end{aligned} \tag{2.15}$$

seen as linear in x_0^2, $x_0 x_1$, ...

This function Φ_2 is a covariant of degree 3 of type $\mathsf{Pol}_3((\mathbb{C}^3)^*, (\mathbb{C}^4)^*)$. The geometric object associated to $\Phi_2(f)$ is a (non–proper) parameterization of the the dual surface to $\mathcal{S}(f)$.

Line whose image spans the same plane.

As already mentioned, any section of $\mathcal{S}(f)$ by some of its tangent planes is a union of two conics. The preimage of each is a straight line in $\mathbb{CP}^2$.

Thus we have the following construction: take a generic line L drawn in $\mathbb{CP}^2$, consider its image in $\mathbb{CP}^3$, this is a conic spanning a tangent plane. The preimage of this plane is made of the original line L, plus another one, L'. The map $L \mapsto L'$ is given by a covariant Φ_3 of type $\mathsf{Pol}_2((\mathbb{C}^3)^*, (\mathbb{C}^3)^*)$. This covariant is defined by the formula

$$\Phi_3 = \begin{vmatrix} a_{00} & a_{01} & a_{02} & 2\,b_{00} & 2\,b_{01} & 2\,b_{02} & 0 \\ a_{10} & a_{11} & a_{12} & 2\,b_{10} & 2\,b_{11} & 2\,b_{12} & 0 \\ a_{20} & a_{21} & a_{22} & 2\,b_{20} & 2\,b_{21} & 2\,b_{22} & 0 \\ a_{30} & a_{31} & a_{32} & 2\,b_{30} & 2\,b_{31} & 2\,b_{32} & 0 \\ \lambda_0 & 0 & 0 & 0 & \lambda_2 & \lambda_1 & x_0 \\ 0 & \lambda_1 & 0 & \lambda_2 & 0 & \lambda_0 & x_1 \\ 0 & 0 & \lambda_2 & \lambda_1 & \lambda_0 & 0 & x_2 \end{vmatrix}. \tag{2.16}$$

It has degree 4.

Implicit equation.

The implicit equation of $\mathcal{S}(f)$ can be obtained as follows. Consider $\Phi_1(f)$ as a cubic polynomial in x:

$$\begin{aligned} \Phi_1 = {} & \ell_{300}(y)x_0^3 + \ell_{030}(y)x_1^3 + \ell_{003}(y)x_2^3 + 3\,\ell_{210}(y)x_0^2 x_1 + 3\,\ell_{201}(y)x_0^2 x_2 \\ & + 3\,\ell_{120}(y)x_1^2 x_0 + 3\,\ell_{021}(y)x_1^2 x_2 + 3\,\ell_{102}(y)x_2^2 x_0 + 3\,\ell_{012}(y)x_2^2 x_1 \\ & + 6\ell_{111}(y)x_0 x_1 x_2. \end{aligned}$$

$$\tag{2.17}$$

Here the coefficients ℓ_{ijk} are linear forms in y, depending polynomially on f. Set

$$\Phi_4 = 6^3 \begin{vmatrix} a_{00} & a_{01} & a_{02} & b_{00} & b_{01} & b_{02} & y_0 \\ a_{10} & a_{11} & a_{12} & b_{10} & b_{11} & b_{12} & y_1 \\ a_{20} & a_{21} & a_{22} & b_{20} & b_{21} & b_{22} & y_2 \\ a_{30} & a_{31} & a_{32} & b_{30} & b_{31} & b_{32} & y_3 \\ \ell_{300} & \ell_{120} & \ell_{102} & \ell_{111} & \ell_{201} & \ell_{210} & 0 \\ \ell_{210} & \ell_{030} & \ell_{012} & \ell_{021} & \ell_{111} & \ell_{120} & 0 \\ \ell_{201} & \ell_{021} & \ell_{003} & \ell_{012} & \ell_{102} & \ell_{111} & 0 \end{vmatrix}. \tag{2.18}$$

Then $\Phi_4(f)$ is an implicit equation of $\mathcal{S}(f)$. And Φ_4 is also a covariant. it has degree 12 and type $\mathsf{Pol}_4(\mathbb{C}^4)$. The attached geometric object is its zero locus, that is merely the surface itself.

This covariant has another property: it vanishes if and only if the parameterization admits a base point (this means that the f_i's have a common zero in $\mathbb{CP}^2$; thus it is revealed to be a resultant).

Formula (2.18) has been proposed in [3]. See [4, 6, 10], for formulas close to this one, and proofs.

Associated Quadric.

One produces a new covariant by the following *contraction* (see [11]) of Φ_1 and Φ_2:

$$\Phi_5 = \frac{1}{6} \sum_{i,j,k} \frac{\partial^3 \Phi_1}{dx_i \, dx_j \, dx_k} \frac{\partial^3 \Phi_2}{d\lambda_i \, d\lambda_j \, d\lambda_k}. \tag{2.19}$$

It has degree 6 and type $\mathsf{Pol}_2(\mathbb{C}^4)$. One finds (by evaluation on the representative of the dense orbit) that $\Phi_5(f) = 0$ is an equation for the Associated Quadric.

Preimage of a point of the Steiner surface.

The map $[f]$ from $\mathbb{CP}^2$ to $\mathbb{CP}^3$ induced by f is birational onto its image $\mathcal{S}(f)$: its inverse is induced by the rational map $[\Phi_6(f)] : \mathbb{CP}^3 \to \mathbb{CP}^2$ where

$$\Phi_6 = 6^3 \begin{vmatrix} a_{00} & a_{01} & a_{02} & b_{00} & b_{01} & b_{02} & 0 \\ a_{10} & a_{11} & a_{12} & b_{10} & b_{11} & b_{12} & 0 \\ a_{20} & a_{21} & a_{22} & b_{20} & b_{21} & b_{22} & 0 \\ a_{30} & a_{31} & a_{32} & b_{30} & b_{31} & b_{32} & 0 \\ \ell_{300} & \ell_{120} & \ell_{102} & \ell_{111} & \ell_{201} & \ell_{210} & \lambda_0 \\ \ell_{210} & \ell_{030} & \ell_{012} & \ell_{021} & \ell_{111} & \ell_{120} & \lambda_1 \\ \ell_{201} & \ell_{021} & \ell_{003} & \ell_{012} & \ell_{102} & \ell_{111} & \lambda_2 \end{vmatrix}. \tag{2.20}$$

This is a covariant of degree 10 and type $\mathsf{Pol}_2(\mathbb{C}^4, \mathbb{C}^3)$.

The dual surface.

Consider the quadratic form $\alpha_0 f_0 + \cdots + \alpha_3 f_3$ and take its discriminant (that is the determinant of its matrix):

$$\Phi_7 = \mathrm{Disc}\,(\alpha_0 f_0 + \alpha_1 f_1 + \alpha_2 f_2 + \alpha_3 f_3)\,. \qquad (2.21)$$

The object obtained this way, Φ_7, is a covariant. It has degree 3 and type $\mathsf{Pol}_3((\mathbb{C}^4)^*)$. The zero locus of $\Phi_7(f)$ is the dual surface to $\mathcal{S}(f)$.

Triple point.

A covariant of degree 9 and type $\mathbb{C}^4 \cong \mathsf{Pol}_1((\mathbb{C}^4)^*)$ is produced by contraction of Φ_7 and Φ_5:

$$\Phi_8 = \sum_{i,j} \frac{\partial^2 \Phi_5}{dy_i\, dy_j} \frac{\partial^2 \Phi_7}{d\alpha_i\, d\alpha_j}. \qquad (2.22)$$

Write

$$\Phi_8(f) = \tau_0 \alpha_0 + \tau_1 \alpha_1 + \tau_2 \alpha_2 + \tau_3 \alpha_3. \qquad (2.23)$$

Then the associated geometric object is a point $(\tau_0 : \tau_1 : \tau_2 : \tau_3)$ of $\mathbb{CP}^3$. One checks that this is exactly the triple point of $\mathcal{S}(f)$.

Discriminant.

By evaluating $\Phi_5(f)$, the equation of the Associated Quadric, at $\Phi_8(f)$, the Triple Point, one gets a scalar:

$$\Delta(f) = \Phi_5(f)(\Phi_8(f)). \qquad (2.24)$$

This object Δ is a homogeneous covariant of degree 24 and type $\mathbb{C}$. Otherwise stated, this is a homogeneous invariant for $SL(3, \mathbb{C}) \times SL(4, \mathbb{C})$. One checks by direct computation that it is irreducible. From this and the existence of a dense orbit, it is not difficult to deduce that Δ is essentially the only invariant. This means that Δ generates the algebra of the invariants under $SL(3, \mathbb{C}) \times SL(4, \mathbb{C})$.

Union of the tropes.

Set

$$\Phi_9 = \Phi_4 + \Phi_5^2. \qquad (2.25)$$

This is a covariant of degree 12 and type $\mathsf{Pol}_4(\mathbb{C}^4)$, and thus $\Phi_9(f)$ represents some quartic surface in $\mathbb{CP}^3$. One checks that this surface is the union of the four tropes.

Trihedron of the double lines.

Remember the classical notion of polar: given an hypersurface of degree $d > 1$ given by an equation $F(z_0, \ldots, z_r) = 0$ and a point $(Z_0 : \cdots : Z_r)$, the polar of the hypersurface and the point is the hypersurface of degree $d - 1$ defined by the equation

$$\sum_i Z_i \frac{\partial F}{dz_i} = 0. \qquad (2.26)$$

Then the polar of $S(f)$ and the triple point $\tau(f)$ has equation $\varphi_5(f) = 0$, where

$$\Phi_{10} = \sum_{i=0}^{3} \tau_i(f) \frac{\partial \Phi_4}{dy_i}. \tag{2.27}$$

(The τ_i's are defined in Equation (2.23).) This way we get a covariant of degree 21 with type $\mathsf{Pol}_3(\mathbb{C}^4)$. One checks that its zero locus in $\mathbb{CP}^3$ is a union of three planes: they are the faces of the trihedron drawn by the singular lines of $S(f)$.

Exceptional Triangle.

Consider the discriminant of Φ_3, quadratic form on $(\mathbb{C}^3)^*$:

$$\Phi_{11} = \frac{1}{8} \begin{vmatrix} \frac{\partial^2 \Phi_3}{d\lambda_0^2} & \frac{\partial^2 \Phi_3}{d\lambda_0 \lambda_1} & \frac{\partial^2 \Phi_3}{d\lambda_0 \lambda_2} \\ \frac{\partial^2 \Phi_3}{d\lambda_0 \lambda_1} & \frac{\partial^2 \Phi_3}{d\lambda_1^2} & \frac{\partial^2 \Phi_3}{d\lambda_1 \lambda_2} \\ \frac{\partial^2 \Phi_3}{d\lambda_0 \lambda_2} & \frac{\partial^2 \Phi_3}{d\lambda_1 \lambda_2} & \frac{\partial^2 \Phi_3}{d\lambda_2^2} \end{vmatrix}. \tag{2.28}$$

This is a covariant of degree 12 and type $\mathsf{Pol}_3(\mathbb{C}^3)$. The zero locus of $\Phi_{11}(f)$ in $\mathbb{CP}^2$ is the Exceptional Triangle[12].

Polar plane Π of the Associated Quadric and the Triple Point.

The polar surface of the Associated Quadric and the Triple Point is a plane, call it Π. It has equation $\Phi_{12}(f) = 0$, where

$$\Phi_{12} = \sum_{i=0}^{3} \tau_i \frac{\partial \Phi_5}{dy_i}. \tag{2.29}$$

This is a covariant of degree 15 and type $(\mathbb{C}^4)^* = \mathsf{Pol}_1(\mathbb{C}^4)$.

Conic, preimage of Π.

By merely substituting y_i with $f_i(x)$ in Φ_{12}, one finds a new covariant Φ_{13}:

$$\Phi_{13}(f)(x) = \Phi_{12}(f)(f(x)). \tag{2.30}$$

The covariant Φ_{13} has degree 16 and type $\mathsf{Pol}_2(\mathbb{C}^3)$. Naturally, $\Phi_{13}(f) = 0$ is the equation of the conic that is the preimage by $[f]$ of the section of $S(f)$ by $\Pi(f)$.

---

[12] The equation obtained this way is of smaller degree than the one obtained by simply substituting the y_i's with the f_i's in Φ_{10}. Actually, this latter is proportional to the square of Φ_{11}.

Dual conic to the preimage of Π.

In [13], parag. 377 is shown a covariant $\Psi(q_1, q_2, \lambda)$ of forms on $\mathbb{C}^3$ (q_1 and q_2 quadratic, λ linear), whose vanishing is a necessary and sufficient condition for the traces of the conics of equations $q_1(x) = 0$ and $q_2(x) = 0$ on the line of equation $\lambda(x) = 0$ to be a harmonic system of points.

Set

$$\Phi_{14} = \sum_{i,j} \frac{\partial \Phi_5}{dy_i dy_j} \Psi(f_i, f_j, \lambda) \tag{2.31}$$

where

$$\lambda(x) = \lambda_0 x_0 + \lambda_1 x_1 + \lambda_2 x_2.$$

Then Φ_{14} is a covariant of degree 8 and type $\mathsf{Pol}_2((\mathbb{C}^3)^*)$. One checks that $\Phi_{14}(f) = 0$ is an equation for the conic of $(\mathbb{CP}^2)^*$ dual to the conic of equation $\Phi_{13}(f) = 0$ of $\mathbb{CP}^2$. Note that the equation we find this way has lower degree than the one obtained by computing the comatrix of the matrix of $\Phi_{13}(f)$ (that would have degree 32).

Quadrilateral, preimage of the four torsal conics.

The union of the four torsal conics is also the intersection between the Associated Quadric (defined by $\Phi_5(f) = 0$) and the Steiner surface. Thus, its preimage is also the preimage of the quadric.

Substitute y_i with f_i in Φ_5, this gives a new covariant Φ_{15} of degree 8 and type $\mathsf{Pol}_4(\mathbb{C}^3)$:

$$\Phi_{15}(f)(x) = \Phi_5(f)(f(x)). \tag{2.32}$$

The zero locus of $\Phi_{15}(f)$ in $\mathbb{CP}^2$ is the quadrilateral, preimage of the union of the torsal conics.

2.6 Application: Equations and inequalities defining the types of Steiner surfaces

We want to recognize the orbits in $\mathcal{U}$, that is the orbits of parameterizations of quartic surfaces (from those of surfaces of smaller degree), and next to discriminate between these orbits.

We consider the first task. After [2] (Proposition 2 and Proposition 5), there are three cases to rule out. The first case is when the parameterization $[f]$ admits a base point (*i.e.* the f_i's have a common zero in $\mathbb{CP}^2$). The second case corresponds to the orbit of the parameterization

$$(x_0^2 : x_1^2 : x_2^2 : x_1 x_2). \tag{2.33}$$

The Zariski closure of its image is a quadric. The third case is the case when the Zariski closure of the image of the parametrization is a plane. A necessary and sufficient condition for being in the first case is the identical vanishing of $\Phi_4(f)$, which translates into a system of polynomial equations of degree 12 in the coefficients of f. The second case is isolated by remarking (by mere evaluation on the representative) that Φ_{11} vanishes identically on the orbit of (2.33), and not on the six orbits of parameterizations giving true quartics. This gives another system of equations of degree 12. The third case is detected by the vanishing of the maximal minors of the 4×6 matrix of the coefficients of the f_i's. This is a system of equations of degree 4.

Now we evaluate the covariants of our collection on the representatives of the six orbits in $\mathcal{U}$, and find that Φ_{14} makes possible the discrimination. Let us explain how: $\Phi_{14}(f)$ is a quadratic form on $\mathbb{R}^3$. Let $M(f)$ be its matrix. Then the *inertia* of $\Phi_{14}(f)$ is the following ordered pair: (number of positive eigenvalues of M, number of negative eigenvalues of $M(f)$). The covariance property of Φ_{14} can be stated as follows:

$$\left(\Phi_{14}(\rho \circ f \circ \theta^{-1})\right)(\lambda) = \det(\theta)^{-6} \det(\rho)^2 \left(\Phi_{14}(f)\right)(\lambda \circ \theta^{-1})$$

Because the powers of the determinants involved in the formulas are even, the inertia of $\Phi_{14}(f)$ takes only one value on each orbit of $\mathcal{F}$ under G. As a consequence, it defines a function on $\mathcal{U}$. Table 2.3 shows its values.

Orbit of $[f]$	inertia of $\Phi_{14}(f)$	equations and inequalities
Ii	$(0,3)$	$A_3 > 0 \wedge A_2 > 0 \wedge A_1 > 0$
Iii	$(2,1)$	$A_3 > 0 \wedge (A_2 \leq 0 \vee A_1 \leq 0)$
Iiii	$(1,2)$	$A_3 < 0$
IIi	$(1,1)$	$A_3 = 0 \wedge A_2 < 0$
IIii	$(0,2)$	$A_3 = 0 \wedge A_2 > 0$
III	$(0,1)$	$A_3 = A_2 = 0$

Table 2.3. Discrimination between the orbits.

It is already an interesting result that the inertia of one quadratic form attached to f is enough to discriminate between the six orbits in $\mathcal{U}$.

Now, we want to go further and define the orbits by equations and inequalities. For this we introduce the characteristic polynomial of $M(f)$:

$$\det(t \cdot I - M(f)) = t^3 + A_1(f) t^2 + A_2(f)t + A_3(f). \tag{2.34}$$

Any condition on the inertia can be translated into equations and inequalities involving the coefficients of $A_i(f)$. The formulas obtained are presented in the last column of Table 2.3. They are obtained trivially, except those for discriminating between inertias $(2,1)$ and $(0,3)$, that makes use of Descartes' law of signs [5].

Note that $A_3(f)$ is a non–trivial invariant of degree 24. Thus it should be proportional to Δ. One finds (by evaluation on the representative of Orbit Ii) that the coefficient of proportionality is positive. Thus in the sign conditions above, we are allowed to substitute A_3 with Δ.

Conclusion

In this paper, we have produced a collection of covariants for quadratic parameterizations of surfaces. We were guided by the geometry of the Steiner surface. In future work, we wish to tackle the problem in a more systematic way: exploiting methods from Invariant Theory, we will try to produce systems of generators for the covariants; or at least to describe all the covariants of low degree.

Acknowledgments

The authors want to thank the readers and anonymous referees, of this paper and of a previous version, for their constructive remarks.

Emmanuel Briand was supported by the European Research Training Network RAAG (Real Algebraic and Analytic Geometry), contract No. HPRN-CT-2001-00271. He wishes to thank the Universidad de Cantabria for its hospitality in 2005.

References

1. Apery, F. (1987). *Models of the Real Projective Plane.* Vieweg.
2. Aries, F., Mourrain, B., and Técourt, J.-P. (2004). Quadratically parameterized surfaces: Algorithms and applications. In *Geometric Modeling and Computing: Seattle 2003*, pages 21–40. Nashboro Press.
3. Aries, F. and Senoussi, R. (1997). Approximation de surfaces paramétriques par des carreaux rationnels du second degré en lancer de rayons. *Revue Internationale de CFAO et d'Informatique Graphique*, 12:627–645.
4. Aries, F. and Senoussi, R. (2001). An implicitization algorithm for rational surfaces with no base points. *Journal of Symbolic Computation*, 31:357–365.
5. Basu, S., Pollack, R., and Roy, M.-F. (2003). *Algorithms in real algebraic geometry*, volume 10 of *Algorithms and Computation in Mathematics*. Springer-Verlag, Berlin.
6. Brill, A. v. (1872). Note über die Gleichung der auf einer Ebene abbildbaren Flächen. *Math. Ann.*, 5:401–403.
7. Coffman, A., Schwartz, A. J., and Stanton, C. (1996). The algebra and geometry of Steiner and other quadratically parametrizable surfaces. *Computer Aided Geometric Design*, 13:257–286.
8. Degen, W. (1996). The types of triangular Bézier surfaces. In Mullineux, G., editor, *The Mathematics of Surfaces IV*, volume 38 of *The Institute of Mathematics and its Applications Conference*, pages 153–171. Clarendon, Oxford.
9. España, M.-L., Baret, F., Aries, F., Chelle, M., Andrieu, B., and Prévot, L. (1999). Modeling maize canopy 3D architecture: Application to reflectance simulation. *Ecological Modeling*, 122:25–43.
10. Jouanolou, J.-P. (1996). Résultants anisotropes : Compléments et applications. *The Electronic Journal of Combinatorics*, 3(2):1–92.
11. Kraft, H. and Procesi, C. (1996). Classical invariant theory, a primer. Lecture Notes. Preliminary version.
12. Popov, V. and Vinberg, E. (1994). *Algebraic Geometry IV*, volume 55 of *Encyclopaedia of Mathematical Science*, chapter Invariant Theory. Springer–Verlag.

13. Salmon, G. (1884). *Traité de Géométrie Analytique à Deux Dimensions, (sections coniques).* Gautier-Villars, Paris.
14. Salmon, G. (1915). *A Treatise on the Analytic Geometry of Three Dimensions.* Longman, Greens and co., fifth edition. Revised by Reginald A. P. Rogers.
15. Sederberg, T. W. and Chen, F. (1995). Implicitization using moving curves and surfaces. *Computer Graphics Siggraph 1995*, 18:301–308.

3

Real Line Arrangements
and Surfaces with Many Real Nodes

Sonja Breske[1], Oliver Labs[2], and Duco van Straten[1]

[1] Institut für Mathematik, Johannes Gutenberg Universität, 55099 Mainz, Germany
Breske@Mathematik.Uni-Mainz.de,
Straten@Mathematik.Uni-Mainz.de
[2] Institut für Mathematik, Universität des Saarlandes, Geb. E2.4, 66123 Saarbrücken,
Germany
Labs@Math.Uni-Sb.de, Mail@OliverLabs.net

Summary. A long standing question is if the maximum number $\mu(d)$ of nodes on a surface of degree d in $\mathbb{P}^3(\mathbb{C})$ can be achieved by a surface defined over the reals which has only real singularities. The currently best known asymptotic lower bound, $\mu(d) \gtrsim \frac{5}{12}d^3$, is provided by Chmutov's construction from 1992 which gives surfaces whose nodes have non-real coordinates.

Using explicit constructions of certain real line arrangements we show that Chmutov's construction can be adapted to give only real singularities. All currently best known constructions which exceed Chmutov's lower bound (i.e., for $d = 3, 4, \ldots, 8, 10, 12$) can also be realized with only real singularities. Thus, our result shows that, up to now, all known lower bounds for $\mu(d)$ can be attained with only real singularities.

We conclude with an application of the theory of real line arrangements which shows that our arrangements are aymptotically the best possible ones for the purpose of constructing surfaces with many nodes. This proves a special case of a conjecture of Chmutov.

3.1 Introduction

A node (or A_1 singularity) in $\mathbb{C}^3$ is a singular point which can be written in the form $x^2 + y^2 + z^2 = 0$ in some local coordinates. We denote by $\mu(d)$ the maximum possible number of nodes on a surface in $\mathbb{P}^3(\mathbb{C})$. The question of determining $\mu(d)$ has a long and rich history. Currently, $\mu(d)$ is only known for $d = 1, 2, \ldots, 6$ (see [1, 12] for sextics and [15] for a recent improvement for septics).

In this paper, we consider the relationship between $\mu(d)$ and the maximum possible number of real nodes on a surface in $\mathbb{P}^3(\mathbb{R})$ which we denote by $\mu^{\mathbb{R}}(d)$. Obviously, $\mu^{\mathbb{R}}(d) \leq \mu(d)$, but do we even have $\mu^{\mathbb{R}}(d) = \mu(d)$? In other words: Can the maximum number of nodes be achieved with real surfaces with real singularities?

The previous question arises naturally because all results in low degree $d \leq 12$ suggest that it could be true (see [1, 8, 9, 15, 19] and table 3.1). But the best known asymptotic lower bound, $\mu(d) \gtrsim \frac{5}{12}d^3$, follows from Chmutov's construction [5] which yields only singularities with non-real coordinates. In this paper, we show that his construction can be adapted to give surfaces with only real singularities (see table 3.1). In the real case we can distinguish between two types of nodes, *conical nodes* $(x^2 + y^2 - z^2 = 0)$ and *solitary points* $(x^2 + y^2 + z^2 = 0)$: Our construction produces only conical nodes.

Notice that in general there are no better real upper bounds for $\mu^{\mathbb{R}}(d)$ known than the well-known complex ones of Miyaoka [17] and Varchenko [20]. But in some cases, for solitary points there exist better bounds via the relation to the zero[th] Betti number. E.g., it has been shown by Nikulin that a K3 surface cannot have more than 10 solitary points (although it can have 16 conical nodes). For quartic surfaces in $\mathbb{P}^3$ this result is probably due to R.W.H.T. Hudson (see [7] for an overview on related results).

We show an upper bound of $\approx \frac{5}{6}d^2$ for the maximum number of real critical points on two levels of real simple line arrangements consisting of d lines; here, *simple* means that no three lines meet in a common point. In [6], Chmutov conjectured this to be the maximum number for all complex plane curves of degree d. He also noticed [5] that such a bound directly implies an upper bound for the number of real nodes of certain surfaces. Our upper bound shows that our examples are asymptotically the best possible real line arrangements for this purpose.

d	1	2	3	4	5	6	7	8	9	10	11	12	13	d
$\mu(d), \mu^{\mathbb{R}}(d) \leq$	0	1	4	16	31	65	104	174	246	360	480	645	832	$\frac{4}{9}d(d-1)^2$
$\mu(d), \mu^{\mathbb{R}}(d) \geq$	0	1	4	16	31	65	99	168	**216**	345	**425**	600	**732**	$\approx \frac{5}{12}\mathbf{d^3}$

Table 3.1. The currently known bounds for the maximum number $\mu(d)$ (resp. $\mu^{\mathbb{R}}(d)$) of nodes on a surface of degree d in $\mathbb{P}^3(\mathbb{C})$ (resp. $\mathbb{P}^3(\mathbb{R})$) are equal. The bold numbers indicate in which cases our result improves the previously known lower bound for $\mu^{\mathbb{R}}(d)$.

3.2 Variants of Chmutov's Surfaces with Many Real Nodes

Let $T_d(z) \in \mathbb{R}[z]$ be the Tchebychev polynomial of degree d with critical values -1 and $+1$ (see fig. 3.2). This can either be defined recursively by $T_0(z) := 1$, $T_1(z) := z$, $T_d(z) := 2 \cdot z \cdot T_{d-1}(z) - T_{d-2}(z)$ for $d \geq 2$, or implicitly by $T_d(\cos(z)) = \cos(dz)$. Chmutov [5] uses them together with the so-called *folding polynomials* $F_d^{A_2}(x, y) \in \mathbb{R}[x, y]$ associated to the root-system A_2 to construct surfaces

$$\text{Chm}_d^{A_2}(x, y, z) := F_d^{A_2}(x, y) + \frac{1}{2}(T_d(z) + 1)$$

with many nodes. These folding polynomials are defined as follows:

$$F_d^{A_2}(x,y) := 2 + \det \begin{pmatrix} x & 1 & 0 & \cdots & & \cdots & 0 \\ 2y & x & & \ddots & & & \vdots \\ 3 & y & & \ddots & \ddots & & \vdots \\ 0 & 1 & \ddots & \ddots & \ddots & \ddots & \vdots \\ \vdots & & \ddots & \ddots & \ddots & \ddots & 0 \\ & & & \ddots & \ddots & \ddots & 1 \\ 0 & \cdots & & 0 & 1 & y & x \end{pmatrix} + \det \begin{pmatrix} y & 1 & 0 & \cdots & & \cdots & 0 \\ 2x & y & & \ddots & & & \vdots \\ 3 & x & & \ddots & \ddots & & \vdots \\ 0 & 1 & \ddots & \ddots & \ddots & \ddots & \vdots \\ \vdots & & \ddots & \ddots & \ddots & \ddots & 0 \\ & & & \ddots & \ddots & \ddots & 1 \\ 0 & \cdots & & 0 & 1 & x & y \end{pmatrix}.$$

$$(3.1)$$

The $F_d^{A_2}(x,y)$ have critical points with only three different critical values: 0, -1, and 8. Thus, the surface $\mathrm{Chm}_d^{A_2}(x,y,z)$ is singular exactly at those points at which the critical values of $F_d^{A_2}(x,y)$ and $\frac{1}{2}(T_d(z)+1)$ sum up to zero (i.e., either both are 0 or the first is -1 and the second is $+1$).

Notice that the plane curve defined by $F_d^{A_2}(x,y)$ consists in fact of d lines. But these are not real lines and the critical points of this folding polynomial also have non-real coordinates. It is natural to ask whether there is a real line arrangement which leads to the same number of critical points. The term *folding polynomials* was introduced in [21] (here we use a slightly different definition). In his article, Withers also described many of their properties, but it was Chmutov [5] who noticed that $F_d^{A_2}(x,y)$ has only few different critical values. In [3], the first author computed the critical points of the other folding polynomials. Among these, there are the following examples which are the real line arrangements we have been looking for (see [3, p. 87–89]):

We define the real folding polynomial $F_{\mathbb{R},d}^{A_2}(x,y) \in \mathbb{R}[x,y]$ associated to the root system A_2 as (see also fig. 3.2)

$$F_{\mathbb{R},d}^{A_2}(x,y) := F_d^{A_2}(x+iy,\ x-iy), \qquad (3.2)$$

where i is the imaginary number. It is easy to see that the $F_{\mathbb{R},d}^{A_2}(x,y)$ have indeed real coefficients. The numbers of critical points are the same as those of $F_d^{A_2}(x,y)$; but now they have real coordinates as the following lemma shows:

Lemma 1. *The real folding polynomial $F_{\mathbb{R},d}^{A_2}(x,y)$ associated to the root system A_2 has $\binom{d}{2}$ real critical points with critical value 0 and*

$$\frac{1}{3}d^2 - d \quad \text{if } d \equiv 0 \mod 3, \qquad \frac{1}{3}d^2 - d + \frac{2}{3} \quad \text{otherwise} \qquad (3.3)$$

real critical points with critical value -1. The other critical points also have real coordinates and have critical value 8.

Proof. We proceed similar to the case discussed in [5], see [3, p. 87–95] for details. To calculate the critical points of the real folding polynomial $F_{\mathbb{R},d}^{A_2}$, we use the map $h^1 : \mathbb{R}^2 \to \mathbb{R}^2$, defined by

$$(u,v) \mapsto \begin{pmatrix} \cos(2\pi(u+v)) + \cos(2\pi u) + \cos(2\pi v) \\ \sin(2\pi(u+v)) - \sin(2\pi u) - \sin(2\pi v) \end{pmatrix}.$$

This is in fact just the real and imaginary part of the first component of the generalized cosine h considered by Withers [21] and Chmutov [5]. It is easy to see that h^1 is a coordinate change if $u - v > 0$, $u + 2v > 0$, and $2u + v < 1$. It transforms the polynomial $F_{\mathbb{R},d}^{A_2}$ into the function $G_d^{A_2} : \mathbb{R}^2 \to \mathbb{R}^2$, defined by

$$G_d^{A_2}(u, v) := F_{\mathbb{R},d}^{A_2}(h^1(u, v)) = 2\cos(2\pi du) + 2\cos(2\pi dv) + 2\cos(2\pi d(u+v)) + 2.$$

The calculation of the critical points of $G_d^{A_2}$ is exactly the same as the one performed in [5]. As the function $G_d^{A_2}$ has $(d-1)^2$ distinct real critical points in the region defined by $u - v > 0$, $u + 2v > 0$, and $2u + v < 1$, the images of these points under the map h^1 are all the critical points of the real folding polynomial $F_{\mathbb{R},d}^{A_2}$ of degree d. In contrast to [5], we get real critical points because h^1 is a map from $\mathbb{R}^2$ into itself.

None of the other root systems yield more critical points on two levels. But as mentioned in [16], the real folding polynomials associated to the root system B_2 give hypersurfaces in $\mathbb{P}^n$, $n \geq 5$, which improve the previously known lower bounds for the maximum number of nodes in higher dimensions slightly (see [16]; [3] gives a detailed discussion of all these folding polynomials and their critical points).

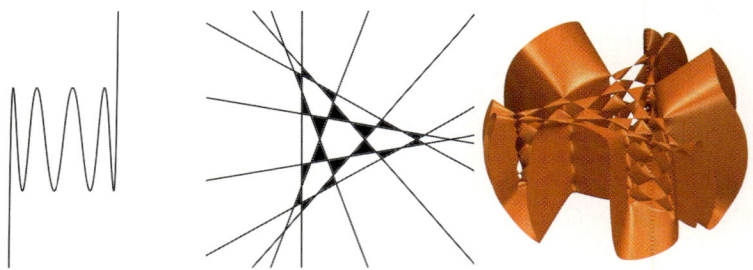

Fig. 3.1. For degree $d = 9$ we show the Tchebychev polynomial $T_9(z)$, the real folding polynomial $F_{\mathbb{R},9}^{A_2}(x, y)$ associated to the root system A_2, and the surface $\mathrm{Chm}_{\mathbb{R},9}^{A_2}(x, y, z)$. The bounded regions in which $F_{\mathbb{R},9}^{A_2}(x, y)$ takes negative values are marked in black.

The lemma immediately gives the following variant of Chmutov's nodal surfaces:

Theorem 2. *Let $d \in \mathbb{N}$. The real projective surface of degree d defined by*

$$\mathrm{Chm}_{\mathbb{R},d}^{A_2}(x, y, z) := F_{\mathbb{R},d}^{A_2}(x, y) + \frac{1}{2}(T_d(z) + 1) \in \mathbb{R}[x, y, z] \qquad (3.4)$$

has the following number of real nodes:

$$
\begin{aligned}
&\tfrac{1}{12}\left(5d^3 - 13d^2 + 12d\right) &&\text{if } d \equiv 0 \mod 6, \\
&\tfrac{1}{12}\left(5d^3 - 13d^2 + 16d - 8\right) &&\text{if } d \equiv 2, 4 \mod 6, \\
&\tfrac{1}{12}\left(5d^3 - 14d^2 + 13d - 4\right) &&\text{if } d \equiv 1, 5 \mod 6, \\
&\tfrac{1}{12}\left(5d^3 - 14d^2 + 9d\right) &&\text{if } d \equiv 3 \mod 6.
\end{aligned}
\qquad (3.5)
$$

These numbers are the same as the numbers of complex nodes of Chmutov's surfaces $\text{Chm}_d^{A_2}(x, y, z)$. To our knowledge, the result gives new lower bounds for the maximum number $\mu^{\mathbb{R}}(d)$ of real singularities on a surface of degree d in $\mathbb{P}^3(\mathbb{R})$ for $d = 9, 11$ and $d \geq 13$, see table 3.1. Notice that all best known lower bounds for $\mu^{\mathbb{R}}(d)$ are attained by surfaces with only conical nodes which is not astonishing in view of the upper bounds for solitary points mentioned in the introduction.

3.3 On Two-Colorings of Real Simple Line Arrangements

The real folding polynomials $F_{\mathbb{R},d}^{A_2}(x, y)$ used in the previous section are in fact *real simple (straight) line arrangements* in $\mathbb{R}^2$, i.e., lines no three of which meet in a point. Such arrangements can be 2-colored in a natural way (see fig. 3.2): We label in black those regions (*cells*) of $\mathbb{R}^2 \setminus \{F_{\mathbb{R},d}^{A_2}(x, y) = 0\}$ in which $F_{\mathbb{R},d}^{A_2}(x, y)$ takes negative values, the others in white. The bounded black regions in fig. 3.2 contain exactly one critical point with critical value -1 each.

Harborth has shown in [11] that the maximum number $M_b(d)$ of black cells in such real simple line arrangements of d lines satisfies:

$$M_b(d) \leq \begin{cases} \frac{1}{3}d^2 + \frac{1}{3}d, & d \text{ odd}, \\ \frac{1}{3}d^2 + \frac{1}{6}d, & d \text{ even}. \end{cases} \tag{3.6}$$

d of these cells are unbounded. This is a purely combinatorial result which is strongly related to the problem of determining the maximum number of triangles in such arrangements which has a long and rich history (see [10]). Notice that this bound is better than the one obtained by Kharlamov using Hodge theory [13]. It is known that the bound (3.6) is exact for infinitely many values of d. The real folding polynomials $F_{\mathbb{R},d}^{A_2}(x, y)$ almost achieve this bound. Moreover, our arrangements have the very special property that all critical points with a negative (resp. positive) critical value have the same critical value -1 (resp. $+8$).

To translate the upper bound on the number of black cells into an upper bound on critical points we use the following lemma:

Lemma 3 (see Lemme 10, 11 in [18]). *Let f be a real simple line arrangement consisting of $d \geq 3$ lines. Then f has exactly $\binom{d-1}{2}$ bounded open cells each of which contains exactly one critical point. Moreover, all the critical points of f are non-degenerate. No unbounded open cell contains a critical point.*

It is easy to prove the lemma, e.g. by counting the number of bounded cells and by observing that each such cell contains at least one critical point. Comparing this with the number $(d-1)^2 - \binom{d}{2} = \binom{d-1}{2}$ of all critical points with non-zero critical values gives the result. Now we can show that our real line arrangements are asymptotically the best possible ones for constructing surfaces with many singularities:

Theorem 4. *The maximum number of critical points with the same non-zero real critical value $0 \neq v \in \mathbb{R}$ of a real simple line arrangement is bounded by $M_b(d) - d$, where d is the number of lines. In particular, the maximum number of critical points on two levels of such an arrangement does not exceed $\binom{d}{2} + M_b(d) - d \approx \frac{5}{6}d^2$.*

Proof. By the preceding lemma, the number of critical points with non-zero critical value equals the number of bounded cells of the real simple line arrangement. The upper bound (3.6) for the maximum number $M_b(d)$ of black cells of a real simple line arrangement now gives the result, because the line arrangement has exactly $\binom{d}{2}$ critical points with critical value 0.

Chmutov showed a much more general result ([4], see [6] for the case of non-degenerate critical points): For a plane curve of degree d the maximum number of critical points on two levels does not exceed $\approx \frac{7}{8}d^2$. In [6], he conjectured $\approx \frac{5}{6}d^2$ to be the actual maximum which is attained by the complex line arrangements $F_d^{A_2}(x,y)$ he used for his construction (and also by the real line arrangement $F_{\mathbb{R},d}^{A_2}(x,y)$). Thus, our theorem 4 is the verification of Chmutov's conjecture in the particular case of real simple line arrangements. As Chmutov remarked in [5], such an upper bound immediately implies an upper bound on the maximum number of nodes on a surface in separated variables:

Corollary 5. *A surface of the form* $p(x,y) + q(z) = 0$ *cannot have more than* $\approx \frac{1}{2}d^2 \cdot \frac{1}{2}d + \frac{1}{3}d^2 \cdot \frac{1}{2}d = \frac{5}{12}d^3$ *nodes if* $p(x,y)$ *is a real simple line arrangement. This number is attained by the surfaces* $\mathrm{Chm}_{\mathbb{R},d}^{A_2}(x,y,z)$ *defined in theorem 2.*

3.4 Concluding Remarks

Comparing our bound from corollary 5 to the upper bound $\approx \frac{5}{12}d^3$ on the zero[th] Betti number (see e.g., [2] or [7]) one is tempted to ask if it is possible to deform our singular surfaces to get examples with many real connected components. But our surfaces $\mathrm{Chm}_{\mathbb{R},d}^{A_2}(x,y,z)$ only contain A_1^- singularities which locally look like a cone ($x^2 + y^2 - z^2 = 0$). When removing the singularities from the zero-set of the surface every connected component contains at least three of the singularities. Thus, the zero[th] Betti number of a small deformation of our surfaces are not larger than $\approx \frac{5}{3\cdot12}d^3$ which is far below the number $\approx \frac{13}{36}d^3$ resulting from Bihan's construction [2].

Conversely, we may ask if it is always possible to move the lines of a simple real line arrangement in such a way that all critical points which have a critical value of the same sign can be chosen to have the same critical value. If this were true then it would be possible to improve our lower bound for the maximum number $\mu^{\mathbb{R}}(d)$ of real nodes on a real surface of degree d slightly because it is known that the upper bounds for the maximum number $M_b(d)$ of black cells are in fact exact for infinitely many d. E.g., in the already cited article [11], Harborth gave an explicit arrangement of 13 straight lines which has $\frac{1}{3}\cdot13^2 + \frac{1}{3}\cdot13 - 13 = 47$ bounded black regions. When regarding this arrangement as a polynomial of degree $d = 13$ it has exactly one critical point with a negative critical value within each of the black regions. If all these negative critical values can be chosen to be the same then such a polynomial will lead to a surface with $\binom{13}{2}\cdot\lceil\frac{13-1}{2}\rceil + 47\cdot\lfloor\frac{13-1}{2}\rfloor = 750 > 732$ nodes. Similarly, such a surface of degree 9 would have $228 > 216$ nodes. In the case of degree 7 the construction would only yield 96 nodes which is less than the number 99 found in [15].

Notice that it is not clear that line arrangements are the best plane curves for our purpose, and we may ask: Is it possible to exceed the number of critical points on two levels of the line arrangements $F_{\mathbb{R},d}^{A_2}(x, y)$ using irreducible curves of higher degrees? Either in the real or in the complex case? This is not true for the real folding polynomials. E.g., those associated to the root system B_2 consist of many ellipses and yield surfaces with fewer singularities (see [3]).

References

1. W. Barth. Two Projective Surfaces with Many Nodes, Admitting the Symmetry of the Icosahedron. *J. Algebraic Geom.*, 5(1):173–186, 1996.
2. F. Bihan. Asymptotic behaviour of Betti numbers of real algebraic surfaces. *Comment. Math. Helv.*, 78:227–244, 2003.
3. S. Breske. Konstruktion von Flächen mit vielen reellen Singularitäten mit Hilfe von Faltungspolynomen. Diploma Thesis. University of Mainz, 2005. Available from [14].
4. S.V. Chmutov. Spectrum and equivariant deformations of critical points. *Uspekhi mat. nauk*, 39(4):113–114, 1984. In Russian.
5. S.V. Chmutov. Examples of Projective Surfaces with Many Singularities. *J. Algebraic Geom.*, 1(2):191–196, 1992.
6. S.V. Chmutov. Extremal distributions of critical points and critical values. In D. T. Lê, K. Saito, and B. Teissier, editors, *Singularity Theory*, pages 192–205, 1995.
7. A. Degtyarev and V. Kharlamov. Topological properties of real algebraic varieties: Du côté de chez Rokhlin. *Russ. Math. Surv.*, 55(4):735–814, 2000.
8. S. Endraß. Flächen mit vielen Doppelpunkten. *DMV-Mitteilungen*, 4(4):17–20, 1995.
9. S. Endraß. A Projective Surface of Degree Eight with 168 Nodes. *J. Algebraic Geom.*, 6(2):325–334, 1997.
10. J.E. Goodman and J. O'Rourke, editors. *Handbook of Computational Geometry*, chapter 5: Pseudoline Arrangements. Chapman & Hall/CRC, 2nd edition, 2004.
11. H. Harborth. Two-Colorings of Simple Arrangements. In *Finite and Infinite Sets*, number 37 in Colloquia Mathematica Societatis János Bolyai, pages 371–378. North-Holland, 1981.
12. D.B. Jaffe and D. Ruberman. A Sextic Surface cannot have 66 Nodes. *J. Algebraic Geom.*, 6(1):151–168, 1997.
13. V. Kharlamov. Overview of topological properties of real algebraic surfaces. math.AG/0502127, 2005.
14. O. Labs. Algebraic Surface Homepage. Information, Images and Tools on Algebraic Surfaces. *www.AlgebraicSurface.net*, 2003.
15. O. Labs. A Septic with 99 Real Nodes. Preprint, math.AG/0409348, to appear in: *Rend. Sem. Mat. Univ. Pad.*, 2004.
16. O. Labs. Dessins D'Enfants and Hypersurfaces in $\mathbb{P}^3$ with many A_j-Singularities. Preprint, math.AG/0505022, 2005.
17. Y. Miyaoka. The Maximal Number of Quotient Singularities on Surfaces with Given Numerical Invariants. *Math. Ann.*, 268:159–171, 1984.
18. A. Ortiz-Rodriguez. Quelques aspects sur la géométrie des surfaces algébriques réelles. *Bull. Sci. Math.*, 127:149–177, 2003.

19. A. Sarti. Pencils of symmetric surfaces in $\mathbb{P}^3$. *J. Algebra*, 246(1):429–452, 2001.
20. A.N. Varchenko. On the Semicontinuity of the Spectrum and an Upper Bound for the Number of Singular Points of a Projective Hypersurface. *J. Soviet Math.*, 270:735–739, 1983.
21. W.D. Withers. Folding Polynomials and Their Dynamics. *Amer. Math. Monthly*, 95:399–413, 1988.

4

Monoid Hypersurfaces

Pål Hermunn Johansen, Magnus Løberg, and Ragni Piene

Centre of Mathematics for Applications and Department of Mathematics
University of Oslo
P. O. Box 1053 Blindern
NO-0316 Oslo, Norway
{hermunn,mags,ragnip}@math.uio.no

Summary. A monoid hypersurface is an irreducible hypersurface of degree d which has a singular point of multiplicity $d-1$. Any monoid hypersurface admits a rational parameterization , hence is of potential interest in computer aided geometric design . We study properties of monoids in general and of monoid surfaces in particular. The main results include a description of the possible real forms of the singularities on a monoid surface other than the $(d-1)$-uple point. These results are applied to the classification of singularities on quartic monoid surfaces , complementing earlier work on the subject.

4.1 Introduction

A monoid hypersurface is an (affine or projective) irreducible algebraic hypersurface which has a singularity of multiplicity one less than the degree of the hypersurface. The presence of such a singular point forces the hypersurface to be rational: there is a rational parameterization given by (the inverse of) the linear projection of the hypersurface from the singular point.

The existence of an explicit rational parameterization makes such hypersurfaces potentially interesting objects in computer aided design. Moreover, since the "space" of monoids of a given degree is much smaller than the space of all hypersurfaces of that degree, one can hope to use monoids efficiently in (approximate or exact) implicitization problems. These were the reasons for considering monoids in the paper [17]. In [12] monoid curves are used to approximate other curves that are close to a monoid curve, and in [13] the same is done for monoid surfaces. In both articles the error of such approximations are analyzed – for each approximation, a bound on the distance from the monoid to the original curve or surface can be computed.

In this article we shall study properties of monoid hypersurfaces and the classification of monoid surfaces with respect to their singularities. Section 4.2 explores properties of monoid hypersurfaces in arbitrary dimension and over an arbitrary base field. Section 4.3 contains results on monoid surfaces, both over arbitrary fields and over $\mathbb{R}$. The last section deals with the classification of monoid surfaces of degree four. Real and complex quartic monoid surfaces were first studied by Rohn [15], who gave a fairly complete description of all possible cases. He also remarked [15, p. 56] that some of his results on quartic monoids hold for monoids of arbitrary degree; in particular, we believe he was aware of many of the results in Section 4.3. Takahashi, Watanabe, and Higuchi [19] classify *complex* quartic monoid surfaces, but do not refer to Rohn. (They cite Jessop [7]; Jessop, however, only treats quartic surfaces with double points and refers to Rohn for the monoid case.) Here we aim at giving a short description of the possible singularities that can occur on quartic monoids, with special emphasis on the real case.

4.2 Basic properties

Let k be a field, let $\bar{k}$ denote its algebraic closure and $\mathbb{P}^n := \mathbb{P}^n_{\bar{k}}$ the projective n-space over $\bar{k}$. Furthermore we define the set of k-rational points $\mathbb{P}^n(k)$ as the set of points that admit representatives $(a_0 : \cdots : a_n)$ with each $a_i \in k$.

For any homogeneous polynomial $F \in \bar{k}[x_0, \ldots, x_n]$ of degree d and point $p = (p_0 : p_1 : \cdots : p_n) \in \mathbb{P}^n$ we can define the multiplicity of $\mathbb{Z}(F)$ at p. We know that $p_r \neq 0$ for some r, so we can assume $p_0 = 1$ and write

$$F = \sum_{i=0}^{d} x_0^{d-i} f_i(x_1 - p_1 x_0, x_2 - p_2 x_0, \ldots, x_n - p_n x_0)$$

where f_i is homogeneous of degree i. Then the multiplicity of $\mathbb{Z}(F)$ at p is defined to be the smallest i such that $f_i \neq 0$.

Let $F \in \bar{k}[x_0, \ldots, x_n]$ be of degree $d \geq 3$. We say that the hypersurface $X = \mathbb{Z}(F) \subset \mathbb{P}^n$ is a *monoid* hypersurface if X is irreducible and has a singular point of multiplicity $d - 1$.

In this article we shall only consider monoids $X = \mathbb{Z}(F)$ where the singular point is k-rational. Modulo a projective transformation of $\mathbb{P}^n$ over k we may – and shall – therefore assume that the singular point is the point $O = (1 : 0 : \cdots : 0)$.

Hence, we shall from now on assume that $X = \mathbb{Z}(F)$, and

$$F = x_0 f_{d-1} + f_d,$$

where $f_i \in k[x_1, \ldots, x_n] \subset k[x_0, \ldots, x_n]$ is homogeneous of degree i and $f_{d-1} \neq 0$. Since F is irreducible, f_d is not identically 0, and f_{d-1} and f_d have no common (non-constant) factors.

The *natural rational parameterization* of the monoid $X = Z(F)$ is the map

$$\theta_F : \mathbb{P}^{n-1} \to \mathbb{P}^n$$

given by

$$\theta_F(a) = (f_d(a) : -f_{d-1}(a)a_1 : \ldots : -f_{d-1}(a)a_n),$$

for $a = (a_1 : \cdots : a_n)$ such that $f_{d-1}(a) \neq 0$ or $f_d(a) \neq 0$.

The set of lines through O form a $\mathbb{P}^{n-1}$. For every $a = (a_1 : \cdots : a_n) \in \mathbb{P}^{n-1}$, the line

$$L_a := \{(s : ta_1 : \ldots : ta_n)|(s : t) \in \mathbb{P}^1\} \tag{4.1}$$

intersects $X = \mathbb{Z}(F)$ with multiplicity at least $d - 1$ in O. If $f_{d-1}(a) \neq 0$ or $f_d(a) \neq 0$, then the line L_a also intersects X in the point

$$\theta_F(a) = (f_d(a) : -f_{d-1}(a)a_1 : \ldots : -f_{d-1}(a)a_n).$$

Hence the natural parameterization is the "inverse" of the projection of X from the point O. Note that θ_F maps $\mathbb{Z}(f_{d-1}) \setminus \mathbb{Z}(f_d)$ to O. The points where the parameterization map is not defined are called base points, and these points are precisely the common zeros of f_{d-1} and f_d. Each such point b corresponds to the line L_b contained in the monoid hypersurface. Additionally, every line of type L_b contained in the monoid hypersurface corresponds to a base point.

Note that $\mathbb{Z}(f_{d-1}) \subset \mathbb{P}^{n-1}$ is the projective tangent cone to X at O, and that $\mathbb{Z}(f_d)$ is the intersection of X with the hyperplane "at infinity" $\mathbb{Z}(x_0)$.

Assume $P \in X$ is another singular point on the monoid X. Then the line L through P and O has intersection multiplicity at least $d - 1 + 2 = d + 1$ with X. Hence, according to Bezout's theorem, L must be contained in X, so that this is only possible if $\dim X \geq 2$.

By taking the partial derivatives of F we can characterize the singular points of X in terms of f_d and f_{d-1}:

Lemma 1. *Let* $\nabla = (\frac{\partial}{\partial x_1}, \ldots, \frac{\partial}{\partial x_n})$ *be the gradient operator.*

(i) *A point* $P = (p_0 : p_1 : \cdots : p_n) \in \mathbb{P}^n$ *is singular on* $\mathbb{Z}(F)$ *if and only if* $f_{d-1}(p_1, \ldots, p_n) = 0$ *and* $p_0 \nabla f_{d-1}(p_1, \ldots, p_n) + \nabla f_d(p_1, \ldots, p_n) = 0$.
(ii) *All singular points of* $\mathbb{Z}(F)$ *are on lines* L_a *where* a *is a base point.*
(iii) *Both* $\mathbb{Z}(f_{d-1})$ *and* $\mathbb{Z}(f_d)$ *are singular in a point* $a \in \mathbb{P}^{n-1}$ *if and only if all points on* L_a *are singular on* X.
(iv) *If not all points on* L_a *are singular, then at most one point other than* O *on* L_a *is singular.*

Proof. (i) follows directly from taking the derivatives of $F = x_0 f_{d-1} + f_d$, and (ii) follows from (i) and the fact that $F(P) = 0$ for any singular point P. Furthermore, a point $(s : ta_1 : \ldots : ta_n)$ on L_a is, by (i), singular if and only if

$$s\nabla f_{d-1}(ta) + \nabla f_d(ta) = t^{d-1}(s\nabla f_{d-1}(a) + t\nabla f_d(a)) = 0.$$

This holds for all $(s : t) \in \mathbb{P}^1$ if and only if $\nabla f_{d-1}(a) = \nabla f_d(a) = 0$. This proves (iii). If either $\nabla f_{d-1}(a)$ or $\nabla f_{d-1}(a)$ are nonzero, the equation above has at most one solution $(s_0 : t_0) \in \mathbb{P}^1$ in addition to $t = 0$, and (iv) follows.

Note that it is possible to construct monoids where $F \in k[x_0, \ldots, x_n]$, but where no points of multiplicity $d - 1$ are k-rational. In that case there must be (at least) two such points, and the line connecting these will be of multiplicity $d - 2$. Furthermore, the natural parameterization will typically not induce a parameterization of the k-rational points from $\mathbb{P}^{n-1}(k)$.

4.3 Monoid surfaces

In the case of a monoid surface, the parameterization has a finite number of base points. From Lemma 1 (ii) we know that all singularities of the monoid other than O, are on lines L_a corresponding to these points. In what follows we will develop the theory for singularities on monoid surfaces — most of these results were probably known to Rohn [15, p. 56].

We start by giving a precise definition of what we shall mean by a monoid surface.

Definition 2. *For an integer $d \geq 3$ and a field k of characteristic 0 the polynomials $f_{d-1} \in k[x_1, x_2, x_3]_{d-1}$ and $f_d \in k[x_1, x_2, x_3]_d$ define a normalized non-degenerate monoid surface $Z(F) \subset \mathbb{P}^3$, where $F = x_0 f_{d-1} + f_d \in k[x_0, x_1, x_2, x_3]$ if the following hold:*

(i) $f_{d-1}, f_d \neq 0$
(ii) $\gcd(f_{d-1}, f_d) = 1$
(iii) *The curves $Z(f_{d-1}) \subset \mathbb{P}^2$ and $Z(f_d) \subset \mathbb{P}^2$ have no common singular point.*

The curves $Z(f_{d-1}) \subset \mathbb{P}^2$ and $Z(f_d) \subset \mathbb{P}^2$ are called respectively the tangent cone *and the* intersection with infinity.

Unless otherwise stated, a surface that satisfies the conditions of Definition 2 shall be referred to simply as a *monoid surface*.

Since we have finitely many base points b and each line L_b contains at most one singular point in addition to O, monoid surfaces will have only finitely many singularities, so all singularities will be isolated. (Note that Rohn included surfaces with nonisolated singularities in his study [15].) We will show that the singularities other than O can be classified by local intersection numbers.

Definition 3. *Let $f, g \in k[x_1, x_2, x_3]$ be nonzero and homogeneous. Assume $p = (p_1 : p_2 : p_3) \in Z(f, g) \subset \mathbb{P}^2$, and define the local intersection number*

$$I_p(f, g) = \lg \frac{\bar{k}[x_1, x_2, x_3]_{m_p}}{(f, g)},$$

where $\bar{k}$ is the algebraic closure of k, $m_p = (p_2 x_1 - p_1 x_2, p_3 x_1 - p_1 x_3, p_3 x_2 - p_2 x_3)$ is the homogeneous ideal of p, and $\lg$ denotes the length of the local ring as a module over itself.

Note that $I_p(f, g) \geq 1$ if and only if $f(p) = g(p) = 0$. When $I_p(f, g) = 1$ we say that f and g intersect transversally at p. The terminology is justified by the following lemma:

Lemma 4. *Let $f, g \in k[x_1, x_2, x_3]$ be nonzero and homogeneous and $p \in Z(f, g)$. Then the following are equivalent:*

(i) $I_p(f, g) > 1$

(ii) f is singular at p, g is singular at p, or $\nabla f(p)$ and $\nabla g(p)$ are nonzero and parallel.

(iii) $s\nabla f(p) + t\nabla g(p) = 0$ for some $(s,t) \neq (0,0)$.

Proof. (ii) is equivalent to (iii) by a simple case study: f is singular at p if and only if (iii) holds for $(s,t) = (1,0)$, g is singular at p if and only if (iii) holds for $(s,t) = (0,1)$, and $\nabla f(p)$ and $\nabla g(p)$ are nonzero and parallel if and only if (iii) holds for some $s, t \neq 0$.

We can assume that $p = (0:0:1)$, so $I_p(f,g) = \lg S$ where

$$ S = \frac{\bar{k}[x_1, x_2, x_3]_{(x_1,x_2)}}{(f,g)}. $$

Furthermore, let $d = \deg f$, $e = \deg g$ and write

$$ f = \sum_{i=1}^{d} f_i x_3^{d-i} \text{ and } g = \sum_{i=1}^{e} g_i x_3^{e-i} $$

where f_i, g_i are homogeneous of degree i.

If f is singular at p, then $f_1 = 0$. Choose $\ell = ax_1 + bx_2$ such that ℓ is not a multiple of g_1. Then ℓ will be a nonzero non-invertible element of S, so the length of S is greater than 1.

We have $\nabla f(p) = (\nabla f_1(p), 0)$ and $\nabla g(p) = (\nabla g_1(p), 0)$. If they are parallel, choose $\ell = ax_0 + bx_1$ such that ℓ is not a multiple of f_1 (or g_1), and argue as above.

Finally assume that f and g intersect transversally at p. We may assume that $f_1 = x_1$ and $g_1 = x_2$. Then $(f,g) = (x_1, x_2)$ as ideals in the local ring $\bar{k}[x_1, x_2, x_3]_{(x_1,x_2)}$. This means that S is isomorphic to the field $\bar{k}(x_3)$. The length of any field is 1, so $I_p(f,g) = \lg S = 1$.

Now we can say which are the lines L_b, with $b \in \mathbb{Z}(f_{d-1}, f_d)$, that contain a singularity other than O:

Lemma 5. *Let f_{d-1} and f_d be as in Definition 2. The line L_b contains a singular point other than O if and only if $\mathbb{Z}(f_{d-1})$ is nonsingular at b and the intersection multiplicity $I_b(f_{d-1}, f_d) > 1$.*

Proof. Let $b = (b_1 : b_2 : b_3)$ and assume that $(b_0 : b_1 : b_2 : b_3)$ is a singular point of $\mathbb{Z}(F)$. Then, by Lemma 1, $f_{d-1}(b_1, b_2, b_3) = f_d(b_1, b_2, b_3) = 0$ and $b_0 \nabla f_{d-1}(b_1, b_2, b_3) + \nabla f_d(b_1, b_2, b_3) = 0$, which implies $I_b(f_{d-1}, f_d) > 1$. Furthermore, if f_{d-1} is singular at b, then the gradient $\nabla f_{d-1}(b_1, b_2, b_3) = 0$, so f_d, too, is singular at b, contrary to our assumptions.

Now assume that $\mathbb{Z}(f_{d-1})$ is nonsingular at $b = (b_1 : b_2 : b_3)$ and the intersection multiplicity $I_b(f_{d-1}, f_d) > 1$. The second assumption implies $f_{d-1}(b_1, b_2, b_3) = f_d(b_1, b_2, b_3) = 0$ and $s\nabla f_{d-1}(b_1, b_2, b_3) = t\nabla f_d(b_1, b_2, b_3)$ for some $(s,t) \neq (0,0)$. Since $\mathbb{Z}(f_{d-1})$ is nonsingular at b, we know that $\nabla f_{d-1}(b_1, b_2, b_3) \neq 0$, so $t \neq 0$. Now $(-s/t : b_1 : b_2 : b_3) \neq (1:0:0:0)$ is a singular point of $\mathbb{Z}(F)$ on the line L_b.

Recall that an A_n singularity is a singularity with normal form $x_1^2 + x_2^2 + x_3^{n+1}$, see [3, p. 184].

Proposition 6. *Let f_{d-1} and f_d be as in Definition 2, and assume $P = (p_0 : p_1 : p_2 : p_3) \neq (1 : 0 : 0 : 0)$ is a singular point of $\mathbb{Z}(F)$ with $I_{(p_1:p_2:p_3)}(f_{d-1}, f_d) = m$. Then P is an A_{m-1} singularity.*

Proof. We may assume that $P = (0 : 0 : 0 : 1)$ and write the local equation

$$g := F(x_0, x_1, x_2, 1) = x_0 f_{d-1}(x_1, x_2, 1) + f_d(x_1, x_2, 1) = \sum_{i=2}^{d} g_i \qquad (4.2)$$

with $g_i \in \bar{k}[x_0, x_1, x_2]$ homogeneous of degree i. Since $\mathbb{Z}(f_{d-1})$ is nonsingular at $0 := (0 : 0 : 1)$, we can assume that the linear term of $f_{d-1}(x_1, x_2, 1)$ is equal to x_1. The quadratic term g_2 of g is then $g_2 = x_0 x_1 + a x_1^2 + b x_1 x_2 + c x_2^2$ for some $a, b, c \in k$. The Hessian matrix of g evaluated at P is

$$H(g)(0,0,0) = H(g_2)(0,0,0) = \begin{pmatrix} 0 & 1 & 0 \\ 1 & 2a & b \\ 0 & b & 2c \end{pmatrix}$$

which has corank 0 when $c \neq 0$ and corank 1 when $c = 0$. By [3, p. 188], P is an A_1 singularity when $c \neq 0$ and an A_n singularity for some n when $c = 0$.

The index n of the singularity is equal to the Milnor number

$$\mu = \dim_{\bar{k}} \frac{\bar{k}[x_0, x_1, x_2]_{(x_0, x_1, x_2)}}{J_g} = \dim_{\bar{k}} \frac{\bar{k}[x_0, x_1, x_2]_{(x_0, x_1, x_2)}}{\left(\frac{\partial g}{\partial x_0}, \frac{\partial g}{\partial x_1}, \frac{\partial g}{\partial x_2} \right)}.$$

We need to show that $\mu = I_0(f_{d-1}, f_d) - 1$. From the definition of the intersection multiplicity, it is not hard to see that

$$I_0(f_{d-1}, f_d) = \dim_{\bar{k}} \frac{\bar{k}[x_1, x_2]_{(x_1, x_2)}}{(f_{d-1}(x_1, x_2, 1), f_d(x_1, x_2, 1))}.$$

The singularity at p is isolated, so the Milnor number is finite. Furthermore, since $\gcd(f_{d-1}, f_d) = 1$, the intersection multiplicity is finite. Therefore both dimensions can be calculated in the completion rings. For the rest of the proof we view f_{d-1} and f_d as elements of the power series rings $\bar{k}[[x_1, x_2]] \subset \bar{k}[[x_0, x_1, x_2]]$, and all calculations are done in these rings.

Since $\mathbb{Z}(f_{d-1})$ is smooth at O, we can write

$$f_{d-1}(x_1, x_2, 1) = (x_1 - \varphi(x_2)) u(x_1, x_2)$$

for some power series $\varphi(x_2)$ and invertible power series $u(x_1, x_2)$. To simplify notation we write $u = u(x_1, x_2) \in \bar{k}[[x_1, x_2]]$.

The Jacobian ideal J_g is generated by the three partial derivatives:

$$\frac{\partial g}{\partial x_0} = (x_1 - \varphi(x_2))\, u$$

$$\frac{\partial g}{\partial x_1} = x_0 \left(u + (x_1 - \varphi(x_2)) \frac{\partial u}{\partial x_1} \right) + \frac{\partial f_d}{\partial x_1}(x_1, x_2)$$

$$\frac{\partial g}{\partial x_2} = x_0 \left(-\varphi'(x_2) u + (x_1 - \varphi(x_2)) \frac{\partial u}{\partial x_2} \right) + \frac{\partial f_d}{\partial x_2}(x_1, x_2)$$

By using the fact that $x_1 - \varphi(x_2) \in \left(\frac{\partial g}{\partial x_0} \right)$ we can write J_g without the symbols $\frac{\partial u}{\partial x_1}$ and $\frac{\partial u}{\partial x_2}$:

$$J_g = \left(x_1 - \varphi(x_2), x_0 u + \frac{\partial f_d}{\partial x_1}(x_1, x_2), -x_0 u \varphi'(x_2) + \frac{\partial f_d}{\partial x_2}(x_1, x_2) \right)$$

To make the following calculations clear, define the polynomials h_i by writing $f_d(x_1, x_2, 1) = \sum_{i=0}^{d} x_1^i h_i(x_2)$. Now

$$J_g = \left(x_1 - \varphi(x_2), x_0 u + \sum_{i=1}^{d} i x_1^{i-1} h_i(x_2), -x_0 u \varphi'(x_2) + \sum_{i=0}^{d} x_1^i h_i'(x_2) \right),$$

so

$$\frac{\bar{k}[[x_0, x_1, x_2]]}{J_g} = \frac{\bar{k}[[x_2]]}{(A(x_2))}$$

where

$$A(x_2) = \varphi'(x_2) \left(\sum_{i=1}^{d} i \varphi(x_2)^{i-1} h_i(x_2) \right) + \left(\sum_{i=0}^{d} \varphi(x_2)^i h_i'(x_2) \right).$$

For the intersection multiplicity we have

$$\frac{\bar{k}[[x_1, x_2]]}{\left(f_{d-1}(x_1, x_2, 1), f_d(x_1, x_2, 1) \right)} = \frac{\bar{k}[[x_1, x_2]]}{\left(x_1 - \varphi(x_2), \sum_{i=0}^{d} x_1^i h_i(x_2) \right)} = \frac{\bar{k}[[x_2]]}{\left(E(x_2) \right)}$$

where $B(x_2) = \sum_{i=0}^{d} \varphi(x_2)^i h_i(x_2)$. Observing that $B'(x_2) = A(x_2)$ gives the result $\mu = I_0(f_{d-1}, f_d) - 1$.

Corollary 7. *A monoid surface of degree d can have at most $\frac{1}{2}d(d-1)$ singularities in addition to O. If this number of singularities is obtained, then all of them will be of type A_1.*

Proof. The sum of all local intersection numbers $I_a(f_{d-1}, f_d)$ is given by Bézout's theorem:

$$\sum_{a \in \mathbb{Z}(f_{d-1}, f_d)} I_a(f_{d-1}, f_d) = d(d-1).$$

The line L_a will contain a singularity other than O only if $I_a(f_{d-1}, f_d) \geq 2$, giving a maximum of $\frac{1}{2}d(d-1)$ singularities in addition to O. Also, if this number is obtained, all local intersection numbers must be exactly 2, so all singularities other than O will be of type A_1.

Both Proposition 6 and Corollary 7 were known to Rohn, who stated these results only in the case $d = 4$, but said they could be generalized to arbitrary d [15, p. 60].

For the rest of the section we will assume $k = \mathbb{R}$. It turns out that we can find a *real* normal form for the singularities other than O. The complex singularities of type A_n come in several real types, with normal forms $x_1^2 \pm x_2^2 \pm x_3^{n+1}$. Varying the $\pm$ gives two types for $n = 1$ and n even, and three types for $n \geq 3$ odd. The real type with normal form $x_1^2 - x_2^2 + x_3^{n+1}$ is called an A_n^- singularity, or *of type A^-*, and is what we find on real monoids:

Proposition 8. *On a real monoid, all singularities other than O are of type A^-.*

Proof. Assume $p = (0 : 0 : 1)$ is a singular point on $\mathbb{Z}(F)$ and set $g = F(x_0, x_1, x_2, 1)$ as in the proof of Proposition 6.

First note that $u^{-1}g = x_0(x_1 - \varphi(x_2)) + f_d(x_1, x_2)u^{-1}$ is an equation for the singularity. We will now prove that $u^{-1}g$ is right equivalent to $\pm(x_0^2 - x_1^2 + x_2^n)$, for some n, by constructing right equivalent functions $u^{-1}g =: g_{(0)} \sim g_{(1)} \sim g_{(2)} \sim g_{(3)} \sim \pm(x_0^2 - x_1^2 + x_2^n)$. Let

$$
\begin{aligned}
g_{(1)}(x_0, x_1, x_2) &= g_{(0)}(x_0, x_1 + \varphi(x_2), x_2) \\
&= x_0 x_1 + f_d(x_1 + \varphi(x_2), x_2)u^{-1}(x_1 + \varphi(x_2), x_2) \\
&= x_0 x_1 + \psi(x_1, x_2)
\end{aligned}
$$

where $\psi(x_1, x_2) \in \mathbb{R}[[x_1, x_2]]$. Write $\psi(x_1, x_2) = x_1 \psi_1(x_1, x_2) + \psi_2(x_2)$ and define

$$
g_{(2)}(x_0, x_1, x_2) = g_{(1)}(x_0 - \psi_1(x_1, x_2), x_1, x_2) = x_0 x_1 + \psi_2(x_2).
$$

The power series $\psi_2(x_2)$ can be written on the form

$$
\psi_2(x_2) = s x_2^n (a_0 + a_1 x_2 + a_2 x_2^2 + \dots)
$$

where $s = \pm 1$ and $a_0 > 0$. We see that $g_{(2)}$ is right equivalent to $g_{(3)} = x_0 x_1 + s x_2^n$ since

$$
g_{(2)}(x_0, x_1, x_2) = g_{(3)}\left(x_0, x_1, x_2 \sqrt[n]{a_0 + a_1 x_2 + a_2 x_2^2 + \dots}\right).
$$

Finally we see that

$$
g_{(4)}(x_0, x_1, x_2) := g_{(3)}(s x_0 - s x_1, x_0 + x_1, x_2) = s(x_0^2 - x_1^2 + x_2^n)
$$

proves that $u^{-1}g$ is right equivalent to $s(x_0^2 - x_1^2 + x_2^n)$ which is an equation for an A_{n-1} singularity with normal form $x_0^2 - x_1^2 + x_2^n$.

Note that for $d = 3$, the singularity at O can be an A_1^+ singularity. This happens for example when $f_2 = x_0^2 + x_1^2 + x_2^2$.

For a *real*, monoid Corollary 7 implies that we can have at most $\frac{1}{2}d(d-1)$ *real* singularities in addition to O. We can show that the bound is sharp by a simple construction:

Example. To construct a monoid with the maximal number of real singularities, it is sufficient to construct two *affine* real curves in the xy-plane defined by equations f_{d-1} and f_d of degrees $d-1$ and d such that the curves intersect in $d(d-1)/2$ points with multiplicity 2. Let $m \in \{d-1, d\}$ be odd and set

$$f_m = \varepsilon - \prod_{i=1}^{m} \left(x \sin\left(\frac{2i\pi}{m}\right) + y \cos\left(\frac{2i\pi}{m}\right) + 1 \right).$$

For $\varepsilon > 0$ sufficiently small there exist at least $\frac{m+1}{2}$ radii $r > 0$, one for each root of the univariate polynomial $f_m|_{x=0}$, such that the circle $x^2 + y^2 - r^2$ intersects f_m in m points with multiplicity 2. Let f_{2d-1-m} be a product of such circles. Now the homogenizations of f_{d-1} and f_d define a monoid surface with $1 + \frac{1}{2}d(d-1)$ singularities. See Figure 4.1.

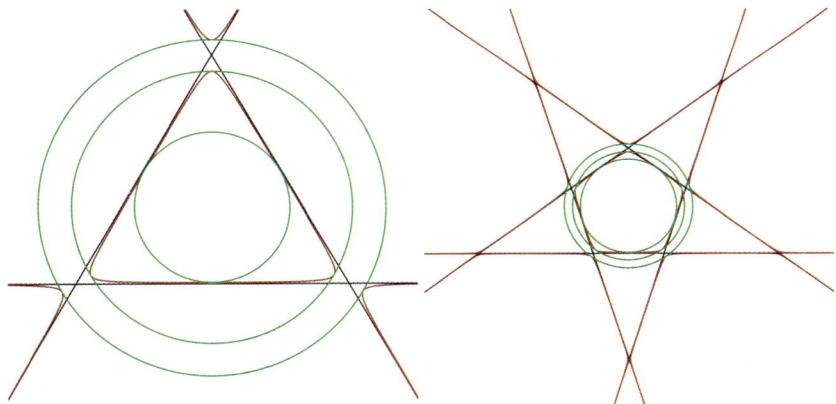

Fig. 4.1. The curves f_m for $m = 3, 5$ and corresponding circles

Proposition 6 and Bezout's theorem imply that the maximal Milnor number of a singularity other than O is $d(d-1) - 1$. The following example shows that this bound can be achieved on a real monoid:

Example. The surface $X \subset \mathbb{P}^3$ defined by $F = x_0(x_1 x_2^{d-2} + x_3^{d-1}) + x_1^d$ has exactly two singular points. The point $(1 : 0 : 0 : 0)$ is a singularity of multiplicity $d-1$ with Milnor number $\mu = (d^2 - 3d + 1)(d-2)$, while the point $(0 : 0 : 1 : 0)$ is an $A_{d(d-1)-1}$ singularity. A picture of this surface for $d = 4$ is shown in Figure 4.2.

Fig. 4.2. The surface defined by $F = x_0(x_1 x_2^{d-2} + x_3^{d-1}) + x_1^d$ for $d = 4$.

4.4 Quartic monoid surfaces

Every cubic surface with isolated singularities is a monoid. Both smooth and singular cubic surfaces have been studied extensively, most notably in [16], where real cubic surfaces and their singularities were classified, and more recently in [18], [4], and [8]. The site [9] contains additional pictures and references.

In this section we shall consider the case $d = 4$. The classification of real and complex *quartic* monoid surfaces was started by Rohn [15]. (In addition to considering the singularities, Rohn studied the existence of lines *not* passing through the triple point, and that of other special curves on the monoid.) In [19], Takahashi, Watanabe, and Higuchi described the singularities of such *complex* surfaces. The monoid singularity of a quartic monoid is minimally elliptic [21], and minimally elliptic singularities have the same complex topological type if and only if their dual graphs are isomorphic [10]. In [10] all possible dual graphs for minimally elliptic singularities are listed, along with example equations.

Using Arnold's notation for the singularities, we use and extend the approach of Takahashi, Watanabe, and Higuchi in [19].

Consider a quartic monoid surface, $X = \mathbb{Z}(F)$, with $F = x_0 f_3 + f_4$. The tangent cone, $\mathbb{Z}(f_3)$, can be of one of nine (complex) types, each needing a separate analysis.

For each type we fix f_3, but any other tangent cone of the same type will be projectively equivalent (over the complex numbers) to this fixed f_3. The nine different types are:

1. Nodal irreducible curve, $f_3 = x_1 x_2 x_3 + x_2^3 + x_3^3$.
2. Cuspidal curve, $f_3 = x_1^3 - x_2^2 x_3$.
3. Conic and a chord, $f_3 = x_3(x_1 x_2 + x_3^2)$
4. Conic and a tangent line, $f_3 = x_3(x_1 x_3 + x_2^2)$.
5. Three general lines, $f_3 = x_1 x_2 x_3$.
6. Three lines meeting in a point, $f_3 = x_2^3 - x_2 x_3^2$

7. A double line and another line, $f_3 = x_2 x_3^2$
8. A triple line $f_3 = x_3^3$
9. A smooth curve, $f_3 = x_1^3 + x_2^3 + x_3^3 + 3a x_0 x_1 x_3$ where $a^3 \neq -1$

To each quartic monoid we can associate, in addition to the type, several integer invariants, all given as intersection numbers. From [19] we know that, for the types 1–3, 5, and 9, these invariants will determine the singularity type of O up to right equivalence. In the other cases the singularity series, as defined by Arnol'd in [1] and [2], is determined by the type of f_3. We shall use, without proof, the results on the singularity type of O due to [19]; however, we shall use the notations of [1] and [2].

We complete the classification begun in [19] by supplying a complete list of the possible singularities occurring on a quartic monoid. In addition, we extend the results to the case of *real* monoids. Our results are summarized in the following theorem.

Theorem 9. *On a quartic monoid surface, singularities other than the monoid point can occur as given in Table 4.1. Moreover, all possibilities are realizable on* real *quartic monoids with a real monoid point, and with the other singularities being real and of type* A^-.

Proof. The invariants listed in the "Invariants and constraints" column are all non-negative integers, and any set of integer values satisfying the equations represents one possible set of invariants, as described above. Then, for each set of invariants, (positive) intersection multiplicities, denoted m_i, m_i' and m_i'', will determine the singularities other than O. The column "Other singularities" give these and the equations they must satisfy. Here we use the notation A_0 for a line L_a on $\mathbb{Z}(F)$ where O is the only singular point.

The analyses of the nine cases share many similarities, and we have chosen not to go into great detail when one aspect of a case differs little from the previous one. We end the section with a discussion on the possible real forms of the tangent cone and how this affects the classification of the *real* quartic monoids.

In all cases, we shall write

$$f_4 = a_1 x_1^4 + a_2 x_1^3 x_2 + a_3 x_1^3 x_3 + a_4 x_1^2 x_2^2 + a_5 x_1^2 x_2 x_3$$
$$+ a_6 x_1^2 x_3^2 + a_7 x_1 x_2^3 + a_8 x_1 x_2^2 x_3 + a_9 x_1 x_2 x_3^2 + a_{10} x_1 x_3^3$$
$$+ a_{11} x_2^4 + a_{12} x_2^3 x_3 + a_{13} x_2^2 x_3^2 + a_{14} x_2 x_3^3 + a_{15} x_3^4$$

and we shall investigate how the coefficients $a_1, \ldots, a_{15}$ are related to the geometry of the monoid.

Case 1. The tangent cone is a nodal irreducible curve, and we can assume

$$f_3(x_1, x_2, x_3) = x_1 x_2 x_3 + x_2^3 + x_3^3.$$

The nodal curve is singular at $(1 : 0 : 0)$. If $f_4(1, 0, 0) \neq 0$, then O is a $T_{3,3,4}$ singularity [19]. We recall that $(1 : 0 : 0)$ cannot be a singular point on $\mathbb{Z}(f_4)$ as

	Triple point	Invariants and constraints	Other singularities
1	$T_{3,3,4}$		$A_{m_i-1}, \sum m_i = 12$
	$T_{3,3,3+m}$	$m = 2, \ldots, 12$	$A_{m_i-1}, \sum m_i = 12 - m$
2	Q_{10}		$A_{m_i-1}, \sum m_i = 12$
	T_{9+m}	$m = 2, 3$	$A_{m_i-1}, \sum m_i = 12 - m$
3	$T_{3,4+r_0,4+r_1}$	$r_0 = \max(j_0, k_0), r_1 = \max(j_1, k_1),$ $j_0 > 0 \leftrightarrow k_0 > 0, \min(j_0, k_0) \leq 1,$ $j_1 > 0 \leftrightarrow k_1 > 0, \min(j_1, k_1) \leq 1$	$A_{m_i-1}, \sum m_i = 4 - k_0 - k_1,$ $A_{m_i'-1}, \sum m_i' = 8 - j_0 - j_1$
4	S series	$j_0 \leq 8, k_0 \leq 4, \min(j_0, k_0) \leq 2,$ $j_0 > 0 \leftrightarrow k_0 > 0, j_1 > 0 \leftrightarrow k_0 > 1$	$A_{m_i-1}, \sum m_i = 4 - k_0,$ $A_{m_i'-1}, \sum m_i' = 8 - j_0$
5	$T_{4+j_k,4+j_l,4+j_m}$	$m_1 + l_1 \leq 4, k_2 + m_2 \leq 4,$ $k_3 + l_3 \leq 4, k_2 > 0 \leftrightarrow k_3 > 0,$ $l_1 > 0 \leftrightarrow l_3 > 0, m_1 > 0 \leftrightarrow m_2 > 0,$ $\min(k_2, k_3) \leq 1, \min(l_1, l_3) \leq 1,$ $\min(m_1, m_2) \leq 1, j_k = \max(k_2, k_3),$ $j_l = \max(l_1, l_3), j_m = \max(m_1, m_2)$	$A_{m_i-1}, \sum m_i = 4 - m_1 - l_1,$ $A_{m_i'-1}, \sum m_i' = 4 - k_2 - m_2,$ $A_{m_i''-1}, \sum m_i'' = 4 - k_3 - l_3$
6	U series	$j_1 > 0 \leftrightarrow j_2 > 0 \leftrightarrow j_3 > 0,$ at most one of $j_1, j_2, j_3 > 1,$ $j_1, j_2, j_3 \leq 4$	$A_{m_i-1}, \sum m_i = 4 - j_1,$ $A_{m_i'-1}, \sum m_i' = 4 - j_2,$ $A_{m_i''-1}, \sum m_i'' = 4 - j_3$
7	V series	$j_0 > 0 \leftrightarrow k_0 > 0, \min(j_0, k_0) \leq 1,$ $j_0 \leq 4, k_0 \leq 4$	$A_{m_i-1}, \sum m_i = 4 - j_0,$
8	V' series		None
9	$P_8 = T_{3,3,3}$		$A_{m_i-1}, \sum m_i = 12$

Table 4.1. Possible configurations of singularities for each case

this would imply a singular line on the monoid, so we assume that either $(1 : 0 : 0) \notin \mathbb{Z}(f_4)$ or $(1 : 0 : 0)$ is a smooth point on $\mathbb{Z}(f_4)$. Let m denote the intersection number $I_{(1:0:0)}(f_3, f_4)$. Since $\mathbb{Z}(f_3)$ is singular at $(1 : 0 : 0)$ we have $m \neq 1$. From [19] we know that O is a $T_{3,3,3+m}$ singularity for $m = 2, \ldots, 12$. Note that some of these complex singularities have two real forms, as illustrated in Figure 4.3.

Bézout's theorem and Proposition 6 limit the possible configurations of singularities on the monoid for each m. Let $\theta(s, t) = (-s^3 - t^3, s^2t, st^2)$. Then the tangent cone $\mathbb{Z}(f_3)$ is parameterized by θ as a map from $\mathbb{P}^1$ to $\mathbb{P}^2$. When we need to compute the intersection numbers between the rational curve $\mathbb{Z}(f_3)$ and the curve $\mathbb{Z}(f_4)$, we can do that by studying the roots of the polynomial $f_4(\theta)$. Expanding the polynomial gives

$$
\begin{aligned}
f_4(\theta)(s, t) = \ & a_1 s^{12} - a_2 s^{11}t + (-a_3 + a_4)s^{10}t^2 + (4a_1 + a_5 - a_7)s^9t^3 \\
& + (-3a_2 + a_6 - a_8 + a_{11})s^8t^4 + (-3a_3 + 2a_4 - a_9 + a_{12})s^7t^5 \\
& + (6a_1 + 2a_5 - a_7 - a_{10} + a_{13})s^6t^6 \\
& + (-3a_2 + 2a_6 - a_8 + a_{14})s^5t^7 + (-3a_3 + a_4 - a_9 + a_{15})s^4t^8 \\
& + (4a_1 + a_5 - a_{10})s^3t^9 + (-a_2 + a_6)s^2t^{10} - a_3st^{11} + a_1t^{12}.
\end{aligned}
$$

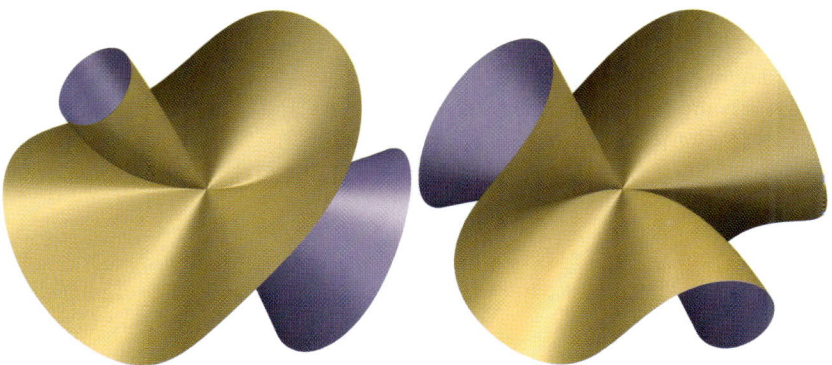

Fig. 4.3. The monoids $\mathbb{Z}(x^3 + y^3 + 5xyz - z^3(x+y))$ and $\mathbb{Z}(x^3 + y^3 + 5xyz - z^3(x-y))$ both have a $T_{3,3,5}$ singularity, but the singularities are not right equivalent over $\mathbb{R}$. (The pictures are generated by the program [5].)

This polynomial will have roots at $(0 : 1)$ and $(1 : 0)$ if and only if $f_4(1,0,0) = a_1 = 0$. When $a_1 = 0$ we may (by symmetry) assume $a_3 \neq 0$, so that $(0 : 1)$ is a simple root and $(1 : 0)$ is a root of multiplicity $m - 1$. Other roots of $f_4(\theta)$ correspond to intersections of $\mathbb{Z}(f_3)$ and $\mathbb{Z}(f_4)$ away from $(1 : 0 : 0)$. The multiplicity m_i of each root is equal to the corresponding intersection multiplicity, giving rise to an A_{m_i-1} singularity if $m_i > 0$, as described by Proposition 6, or a line $L_a \subset \mathbb{Z}(F)$ with O as the only singular point if $m_i = 1$.

The polynomial $f_4(\theta)$ defines a linear map from the coefficient space k^{15} of f_4 to the space of homogeneous polynomials of degree 12 in s and t. By elementary linear algebra, we see that the image of this map is the set of polynomials of the form

$$b_0 s^{12} + b_1 s^{11}t + b_2 s^{10}t^2 + \cdots + b_{12}t^{12}$$

where $b_0 = b_{12}$. The kernel of the map corresponds to the set of polynomials of the form ℓf_3 where ℓ is a linear form. This means that $f_4(\theta) \equiv 0$ if and only if f_3 is a factor in f_4, making $\mathbb{Z}(F)$ reducible and not a monoid.

For every $m = 0, 2, 3, 4, \ldots, 12$ we can select r parameter points

$$p_1, \ldots, p_r \in \mathbb{P}^1 \setminus \{(0 : 1), (1 : 0)\}$$

and positive multiplicities $m_1, \ldots, m_r$ with $m_1 + \cdots + m_r = 12 - m$ and try to describe the polynomials f_4 such that $f_4(\theta)$ has a root of multiplicity m_i at p_i for each $i = 1, \ldots, r$.

Still assuming $a_3 \neq 0$ whenever $a_1 = 0$, any such choice of parameter points $p_1, \ldots, p_r$ and multiplicities $m_1, \ldots, m_r$ corresponds to a polynomial $q = b_0 s^{12} + b_1 s^{11}t + \cdots + b_{12}t^{12}$ that is, up to a nonzero constant, uniquely determined.

Now, q is equal to $f_4(\theta)$ for some f_4 if and only if $b_0 = b_{12}$. If $m \geq 2$, then q contains a factor st^{m-1}, so $b_0 = b_{12} = 0$, giving $q = f_4(\theta)$ for some f_4. In fact, when $m \geq 2$ any choice of $p_1, \ldots, p_r$ and $m_1, \ldots, m_r$ with $m_1 - \cdots + m_r = 12 - m$ corresponds to a four dimensional space of equations f_4 that gives this set of roots and multiplicities in $f_4(\theta)$. If f_4' is one such f_4, then any other is of the form $\lambda f_4' + \ell f_3$ for some constant $\lambda \neq 0$ and linear form ℓ. All of these give monoids that are projectively equivalent.

When $m = 0$, we write $p_i = (\alpha_i : \beta_i)$ for $i = 1, \ldots, r$. The condition $b_0 = b_{12}$ on the coefficients of q translates to

$$\alpha_1^{m_1} \cdots \alpha_r^{m_r} = \beta_1^{m_1} \cdots \beta_r^{m_r}. \tag{4.3}$$

This means that any choice of parameter points $(\alpha_1 : \beta_1), \ldots, (\alpha_r : \beta_r)$ and multiplicities $m_1, \ldots, m_r$ with $m_1 + \cdots + m_r = 12$ that satisfy condition (4.3) corresponds to a four dimensional family $\lambda f_4' + \ell f_3$, giving a unique monoid up to projective equivalence.

For example, we can have an A_{11} singularity only if $f_4(\theta)$ is of the form $(\alpha s - \beta t)^{12}$. Condition (4.3) implies that this can only happen for 12 parameter points, all of the form $(1 : \omega)$, where $\omega^{12} = 1$. Each such parameter point $(1 : \omega)$ corresponds to a monoid uniquely determined up to projective equivalence. However, since there are six projective transformations of the plane that maps $\mathbb{Z}(f_3)$ onto itself, this correspondence is not one to one. If $\omega_1^{12} = \omega_2^{12} = 1$, then ω_1 and ω_2 will correspond to projectively equivalent monoids if and only if $\omega_1^3 = \omega_2^3$ or $\omega_1^3 \omega_2^3 = 1$. This means that there are three different quartic monoids with one $T_{3,3,4}$ singularity and one A_{11} singularity. One corresponds to those ω where $\omega^3 = 1$, one to those ω where $\omega^3 = -1$, and one to those ω where $\omega^6 = -1$. The first two of these have real representatives, $\omega = \pm 1$.

It easy to see that for any set of multiplicities $m_1 + \cdots + m_r = 12$, we can find real points $p_1, \ldots, p_r$ such that condition (4.3) is satisfied. This completely classifies the possible configurations of singularities when f_3 is an irreducible nodal curve.

Case 2. The tangent cone is a cuspidal curve, and we can assume $f_3(x_1, x_2, x_3) = x_1^3 - x_2^2 x_3$. The cuspidal curve is singular at $(0 : 0 : 1)$ and can be parameterized by θ as a map from $\mathbb{P}^1$ to $\mathbb{P}^2$ where $\theta(s, t) = (s^2 t, s^3, t^3)$. The intersection numbers are determined by the degree 12 polynomial $f_4(\theta)$. As in the previous case, $f_4(\theta) \equiv 0$ if and only if f_3 is a factor of f_4, and we will assume this is not the case. The multiplicity m of the factor s in $f_4(\theta)$ determines the type of singularity at O. If $m = 0$ (no factor s), then O is a Q_{10} singularity. If $m = 2$ or $m = 3$, then O is of type Q_{9+m}. If $m > 3$, then $(0 : 0 : 1)$ is a singular point on $\mathbb{Z}(f_4)$, so the monoid has a singular line and is not considered in this article. Also, $m = 1$ is not possible, since $f_4(\theta(s, t)) = f_4(s^2 t, s^3, t^3)$ cannot contain st^{11} as a factor.

For each $m = 0, 2, 3$ we can analyze the possible configurations of other singularities on the monoid. Similarly to the previous case, any choice of parameter points $p_1, \ldots, p_r \in \mathbb{P}^1 \setminus \{(0 : 1)\}$ and positive multiplicities $m_1, \ldots, m_r$ with $\sum m_i = 12 - m$ corresponds, up to a nonzero constant, to a unique degree 12 polynomial q.

When $m = 2$ or $m = 3$, for any choice of parameter values and associated multiplicities, we can find a four dimensional family $f_4 = \lambda f_4' + \ell f_3$ with the prescribed roots in $f_4(\theta)$. As before, the family gives projectively equivalent monoids.

When $m = 0$, one condition must be satisfied for q to be of the form $f_4(\theta)$, namely $b_{11} = 0$, where b_{11} is the coefficient of st^{11} in q.

For example, we can have an A_{11} singularity only if q is of the form $(\alpha s - \beta t)^{12}$. The condition $b_{11} = 0$ implies that either $q = \lambda s^{12}$ or $q = \lambda t^{12}$. The first case gives a surface with a singular line, while the other gives a monoid with an A_{11} singularity (see Figure 4.2). The line from O to the A_{11} singularity corresponds to the inflection point of $\mathbb{Z}(f_3)$.

For any set of multiplicities $m_1, \ldots, m_r$ with $m_1 + \cdots + m_r = 12$, it is not hard to see that there exist real points $p_1, \ldots, p_r$ such that the condition $b_{11} = 0$ is satisfied. It suffices to take $p_i = (\alpha_i : 1)$, with $\sum m_i \alpha_i = 0$ (the condition corresponding to $b_{11} = 0$). This completely classifies the possible configurations of singularities when f_3 is a cuspidal curve.

Case 3. The tangent cone is the product of a conic and a line that is not tangent to the conic, and we can assume $f_3 = x_3(x_1 x_2 + x_3^2)$. Then $\mathbb{Z}(f_3)$ is singular at $(1 : 0 : 0)$ and $(0 : 1 : 0)$, the intersections of the conic $\mathbb{Z}(x_1 x_2 + x_3^2)$ and the line $\mathbb{Z}(x_3)$. For each f_4 we can associate four integers:

$$j_0 := \mathrm{I}_{(1:0:0)}(x_1 x_2 + x_3^2, f_4), \quad k_0 := \mathrm{I}_{(1:0:0)}(x_3, f_4),$$
$$j_1 := \mathrm{I}_{(0:1:0)}(x_1 x_2 + x_3^2, f_4), \quad k_1 := \mathrm{I}_{(0:1:0)}(x_3, f_4).$$

We see that $k_0 > 0 \Leftrightarrow f_4(1 : 0 : 0) = 0 \Leftrightarrow j_0 > 0$, and that $\mathbb{Z}(f_4)$ is singular at $(1 : 0 : 0)$ if and only if k_0 and j_0 both are bigger than one. These cases imply a singular line on the monoid, and are not considered in this article. The same holds for k_1, j_1 and the point $(0 : 1 : 0)$.

Define $r_i = \max(j_i, k_i)$ for $i = 1, 2$. Then, by [19], O will be a singularity of type $T_{3,4+r_0,4+r_1}$ if $r_0 \le r_1$, or of type $T_{3,4+r_1,4+r_0}$ if $r_0 \ge r_1$.

We can parameterize the line $\mathbb{Z}(x_3)$ by θ_1 where $\theta_1(s, t) = (s, t, 0)$, and the conic $\mathbb{Z}(x_1 x_2 + x_3^2)$ by θ_2 where $\theta_2(s, t) = (s^2, -t^2, st)$. Similarly to the previous cases, roots of $f_4(\theta_1)$ correspond to intersections between $\mathbb{Z}(f_4)$ and the line $\mathbb{Z}(x_3)$, while roots of $f_4(\theta_2)$ correspond to intersections between $\mathbb{Z}(f_4)$ and the conic $\mathbb{Z}(x_1 x_3 + x_3^2)$.

For any legal values of of j_0, j_1, k_0 and k_1, parameter points

$$(\alpha_1 : \beta_1), \ldots, (\alpha_{m_r} : \beta_{m_r}) \in \mathbb{P}^1 \setminus \{(0 : 1), (1 : 0)\},$$

with multiplicities $m_1, \ldots, m_r$ such that $m_1 + \cdots + m_r = 4 - k_0 - k_1$, and parameter points

$$(\alpha_1' : \beta_1'), \ldots, (\alpha_{m_{r'}}' : \beta_{m_{r'}}') \in \mathbb{P}^1 \setminus \{(0 : 1), (1 : 0)\},$$

with multiplicities $m_1', \ldots, m_{r'}'$ such that $m_1' + \cdots + m_{r'}' = 8 - j_0 - j_1$, we can fix polynomials q_1 and q_2 such that

- q_1 is nonzero, of degree 4, and has factors s^{k_1}, t^{k_0} and $(\beta_i s - \alpha_i t)^{m_i}$ for $i = 1, \ldots, r$,
- q_2 is nonzero, of degree 8, and has factors s^{j_1}, t^{j_0} and $(\beta_i' s - \alpha_i' t)^{m_i'}$ for $i = 1, \ldots, r'$.

Now q_1 and q_2 are determined up to multiplication by nonzero constants. Write $q_1 = b_0 s^4 + \cdots + b_4 t^4$ and $q_2 = c_0 s^8 + \cdots + c_8 t^8$.

The classification of singularities on the monoid consists of describing the conditions on the parameter points and nonzero constants λ_1 and λ_2 for the pair $(\lambda_1 q_1, \lambda_2 q_2)$ to be on the form $(f_4(\theta_1), f_4(\theta_2))$ for some f_4.

Similarly to the previous cases, $f_4(\theta_1) \equiv 0$ if and only if x_3 is a factor in f_4 and $f_4(\theta_2) \equiv 0$ if and only if $x_1 x_2 + x_3^2$ is a factor in f_4. Since $f_3 = x_3(x_1 x_2 + x_3^2)$, both cases will make the monoid reducible, so we only consider $\lambda_1, \lambda_2 \neq 0$.

We use linear algebra to study the relationship between the coefficients $a_1 \ldots a_{15}$ of f_4 and the polynomials q_1 and q_2. We find $(\lambda_1 q_1, \lambda_2 q_2)$ to be of the form $(f_4(\theta_1), f_4(\theta_2))$ if and only if $\lambda_1 b_0 = \lambda_2 c_0$ and $\lambda_1 b_4 = \lambda_2 c_8$. Furthermore, the pair $(\lambda_1 q_1, \lambda_2 q_2)$ will fix f_4 modulo f_3. Since f_4 and λf_4 correspond to projectively equivalent monoids for any $\lambda \neq 0$, it is the ratio λ_1/λ_2, and not λ_1 and λ_2, that is important.

Recall that $k_0 > 0 \Leftrightarrow j_0 > 0$ and $k_1 > 0 \Leftrightarrow j_1 > 0$. If $k_0 > 0$ and $k_1 > 0$, then $b_0 = c_0 = b_4 = c_8 = 0$, so for any $\lambda_1, \lambda_2 \neq 0$ we have $(\lambda_1 q_1, \lambda_2 q_2) = (f_4(\theta_1), f_4(\theta_2))$ for some f_4. Varying λ_1/λ_2 will give a one-parameter family of monoids for each choice of multiplicities and parameter points.

If $k_0 = 0$ and $k_1 > 0$, then $b_0 = c_0 = 0$. The condition $\lambda_1 b_4 = \lambda_2 c_8$ implies $\lambda_1/\lambda_2 = c_8/b_4$. This means that any choice of multiplicities and parameter points will give a unique monoid up to projective equivalence. The same goes for the case where $k_0 > 0$ and $k_1 = 0$.

Finally, consider the case where $k_0 = k_1 = 0$. For $(\lambda_1 q_1, \lambda_2 q_2)$ to be of the form $(f_4(\theta_1), f_4(\theta_2))$ we must have $\lambda_1/\lambda_2 = c_8/b_4 = c_0/b_0$. This translates into a condition on the parameter points, namely

$$\frac{(\beta_1')^{m_1'} \cdots (\beta_{r'}')^{m_{r'}'}}{\beta_1^{m_1} \cdots \beta_r^{m_r}} = \frac{(\alpha_1')^{m_1'} \cdots (\alpha_{r'}')^{m_{r'}'}}{\alpha_1^{m_1} \cdots \alpha_r^{m_r}}. \tag{4.4}$$

In other words, if condition (4.4) holds, we have a unique monoid up to projective equivalence.

It is easy to see that for any choice of multiplicities, it is possible to find real parameter points such that condition (4.4) is satisfied. This completes the classification of possible singularities when the tangent cone is a conic plus a chordal line.

Case 4. The tangent cone is the product of a conic and a line tangent to the conic, and we can assume $f_3 = x_3(x_1 x_3 + x_2^2)$. Now $\mathbb{Z}(f_3)$ is singular at $(1 : 0 : 0)$. For each f_4 we can associate two integers

$$j_0 := \mathrm{I}_{(1:0:0)}(x_1 x_3 + x_2^2, f_4) \qquad \text{and} \qquad k_0 := \mathrm{I}_{(1:0:0)}(x_3, f_4).$$

We have $j_0 > 0 \Leftrightarrow k_0 > 0$, $j_0 > 1 \Leftrightarrow k_0 > 1$. Furthermore, j_0 and k_0 are both greater than 2 if and only if $\mathbb{Z}(f_4)$ is singular at $(1 : 0 : 0)$, a case we have excluded. The singularity at O will be of the S series, from [1], [2].

We can parameterize the conic $\mathbb{Z}(x_1 x_3 + x_2^2)$ by θ_2 and the line $\mathbb{Z}(x_3)$ by θ_1 where $\theta_2(s, t) = (s^2, st, -t^2)$ and $\theta_1(s, t) = (s, t, 0)$. As in the previous case, the monoid is reducible if and only if $f_4(\theta_1) \equiv 0$ or $f_4(\theta_2) \equiv 0$. Consider two nonzero polynomials

$$q_1 = b_0 s^4 + b_1 s^3 t + b_2 s^2 t^2 + b_3 s t^3 + b_4 t^4$$
$$q_2 = c_0 s^8 + c_1 s^7 t + \cdots + c_7 s t^7 + c_8 t^8.$$

Now $(\lambda_1 q_1, \lambda_2 q_2) = (f_4(\theta_1), f_4(\theta_2))$ for some f_4 if and only if $\lambda_1 b_0 = \lambda_2 c_0$ and $\lambda_1 b_1 = \lambda_2 c_1$. As before, only the cases where $\lambda_1, \lambda_2 \neq 0$ are interesting. We see that $(\lambda_1 q_1, \lambda_2 q_2) = (f_4(\theta_1), f_4(\theta_2))$ for some $\lambda_1, \lambda_2 \neq 0$ if and only if the following hold:

- $b_0 = 0 \leftrightarrow c_0 = 0$ and $b_1 = 0 \leftrightarrow c_1 = 0$
- $b_0 c_1 = b_1 c_0$.

The classification of other singularities (than O) is very similar to the previous case. Roots of $f_4(\theta_1)$ and $f_4(\theta_2)$ away from $(1 : 0)$ correspond to intersections of $Z(f_3)$ and $Z(f_4)$ away from the singular point of $Z(f_3)$, and when one such intersection is multiple, there is a corresponding singularity on the monoid.

Now assume $(\lambda_1 q_1, \lambda_2 q_2) = (f_4(\theta_1), f_4(\theta_2))$ for some $\lambda_1, \lambda_2 \neq 0$ and some f_4. If $b_0 \neq 0$ (equivalent to $c_0 \neq 0$) then $j_0 = k_0 = 0$ and $\lambda_1/\lambda_2 = c_0/b_0$. If $b_0 = c_0 = 0$ and $b_1 \neq 0$ (equivalent to $c_1 \neq 0$), then $j_0 = k_0 = 1$, and $\lambda_1/\lambda_2 = c_1/b_1$. If $b_0 = b_1 = c_0 = c_1 = 0$, then $j_0, k_0 > 1$ and any value of λ_1/λ_2 will give $(\lambda_1 q_1, \lambda_2 q_2)$ of the form $(f_4(\theta_1), f_4(\theta_2))$ for some f_4. Thus we get a one-dimensional family of monoids for this choice of q_1 and q_2.

Now consider the possible configurations of other singularities on the monoid. Assume that $j'_0 \leq 8$ and $k'_0 \leq 4$ are nonnegative integers such that $j_0 > 0 \leftrightarrow k_0 > 0$ and $j_0 > 1 \leftrightarrow k_0 > 1$. For any set of multiplicities $m_1, \ldots, m_r$ with $m_1 + \cdots + m_r = 4 - k'_0$ and $m'_1, \ldots, m'_{r'}$ with $m'_1 + \cdots + m'_{r'} = 8 - j'_0$. there exists a polynomial f_4 with real coefficients such that $f_4(\theta_1)$ has real roots away from $(1 : 0)$ with multiplicities $m_1, \ldots, m_r$, and $f_4(\theta_2)$ has real roots away from $(1 : 0)$ with multiplicities $m'_1, \ldots, m'_{r'}$. Furthermore, for this f_4 we have $k_0 = k'_0$ and $j_0 = j'_0$. Proposition 6 will give the singularities that occur in addition to O.

This completes the classification of the singularities on a quartic monoid (other than O) when the tangent cone is a conic plus a tangent.

Case 5. The tangent cone is three general lines, and we assume $f_3 = x_1 x_2 x_3$. For each f_4 we associate six integers,

$$k_2 := I_{(1:0:0)}(f_4, x_2), \quad l_1 := I_{(0:1:0)}(f_4, x_1), \quad m_1 := I_{(0:0:1)}(f_4, x_1),$$
$$k_3 := I_{(1:0:0)}(f_4, x_3), \quad l_3 := I_{(0:1:0)}(f_4, x_3), \quad m_2 := I_{(0:0:1)}(f_4, x_2).$$

Now $k_2 > 0 \Leftrightarrow k_3 > 0$, $l_1 > 0 \Leftrightarrow l_3 > 0$, and $m_1 > 0 \Leftrightarrow m_2 > 0$. If both k_2 and k_3 are greater than 1, then the monoid has a singular line, a case we have excluded. The same goes for the pairs (l_1, l_3) and (m_1, m_2).

When the monoid does not have a singular line, we define $j_k = \max(k_2, k_3)$, $j_l = \max(l_1, l_3)$ and $j_m = \max(m_1, m_2)$. If $j_k \leq j_l \leq j_m$, then [19] gives that O is a $T_{4+j_k, 4+j_l, 4+j_m}$ singularity.

The three lines $\mathbb{Z}(x_1)$, $\mathbb{Z}(x_2)$ and $\mathbb{Z}(x_3)$ are parameterized by θ_1, θ_2 and θ_3 where $\theta_1(s,t) = (0,s,t)$, $\theta_2(s,t) = (s,0,t)$ and $\theta_3(s,t) = (s,t,0)$. Roots of the polynomial $f_4(\theta_i)$ away from $(1:0)$ and $(0:1)$ correspond to intersections between $\mathbb{Z}(f_4)$ and $\mathbb{Z}(x_i)$ away from the singular points of $\mathbb{Z}(f_3)$.

As before, we are only interested in the cases where none of $f_4(\theta_i) \equiv 0$ for $i = 1, 2, 3$, as this would make the monoid reducible.

For the study of other singularities on the monoid we consider nonzero polynomials

$$q_1 = b_0 s^4 + b_1 s^3 t + b_2 s^2 t^2 + b_3 s t^3 + b_4 t^4,$$
$$q_2 = c_0 s^4 + c_1 s^3 t + c_2 s^2 t^2 + c_3 s t^3 + c_4 t^4,$$
$$q_3 = d_0 s^4 + d_1 s^3 t + d_2 s^2 t^2 + d_3 s t^3 + d_4 t^4.$$

Linear algebra shows that $(\lambda_1 q_1, \lambda_2 q_2, \lambda_3 q_3) = (f_4(\theta_1), f_4(\theta_2), f_4(\theta_3))$ for some f_4 if and only if $\lambda_1 b_0 = \lambda_3 d_4$, $\lambda_1 b_4 = \lambda_2 c_4$, and $\lambda_2 c_0 = \lambda_3 d_0$. A simple analysis shows the following: There exist $\lambda_1, \lambda_2, \lambda_3 \neq 0$ such that

$$(\lambda_1 q_1, \lambda_2 q_2, \lambda_3 q_3) = (f_4(\theta_1), f_4(\theta_2), f_4(\theta_3))$$

for some f_4, and such that $\mathbb{Z}(f_4)$ and $\mathbb{Z}(f_3)$ have no common singular point if and only if all of the following hold:

- $b_0 = 0 \leftrightarrow d_4 = 0$ and $b_0 = d_4 = 0 \rightarrow (b_1 \neq 0$ or $d_3 \neq 0)$,
- $b_4 = 0 \leftrightarrow c_4 = 0$ and $b_4 = c_4 = 0 \rightarrow (b_3 \neq 0$ or $c_3 \neq 0)$,
- $c_0 = 4 \leftrightarrow d_0 = 0$ and $c_0 = d_0 = 0 \rightarrow (c_1 \neq 0$ or $d_1 \neq 0)$,
- $b_0 c_4 d_0 = b_4 c_0 d_4$.

Similarly to the previous cases we can classify the possible configurations of other singularities by varying the multiplicities of the roots of the polynomials q_1, q_2 and q_3. Only the multiplicities of the roots $(0:1)$ and $(1:0)$ affect the first three bullet points above. Then, for any set of multiplicities of the rest of the roots, we can find q_1, q_2 and q_3 such that the last bullet point is satisfied. This completes the classification when $\mathbb{Z}(f_3)$ is the product of three general lines.

Case 6. The tangent cone is three lines meeting in a point, and we can assume that $f_3 = x_2^3 - x_2 x_3^2$. We write $f_3 = \ell_1 \ell_2 \ell_3$ where $\ell_1 = x_2$, $\ell_2 = x_2 - x_3$ and $\ell_3 = x_2 + x_3$, representing the three lines going through the singular point $(1:0:0)$. For each f_4 we associate three integers j_1, j_2 and j_3 defined as the intersection numbers $j_i = \mathrm{I}_{(1:0:0)}(f_4, \ell_i)$. We see that $j_1 = 0 \Leftrightarrow j_2 = 0 \Leftrightarrow j_3 = 0$, and that $\mathbb{Z}(f_4)$ is singular at $(1:0:0)$ if and only if two of the integers j_1, j_2, j_3 are greater then one. (Then all of them will be greater than one.) The singularity will be of the U series [1], [2].

The three lines $\mathbb{Z}(\ell_1)$, $\mathbb{Z}(\ell_2)$ and $\mathbb{Z}(\ell_3)$ can be parameterized by θ_1, θ_2, and θ_3 where $\theta_1(s,t) = (s,0,t)$, $\theta_2(s,t) = (s,t,t)$ and $\theta_2(s,t) = (s,t,-t)$.

For the study of other singularities on the monoid we consider nonzero polynomials

$$q_1 = b_0 s^4 + b_1 s^3 t + b_2 s^2 t^2 + b_3 s t^3 + b_4 t^4,$$
$$q_2 = c_0 s^4 + c_1 s^3 t + c_2 s^2 t^2 + c_3 s t^3 + c_4 t^4,$$
$$q_3 = d_0 s^4 + d_1 s^3 t + d_2 s^2 t^2 + d_3 s t^3 + d_4 t^4.$$

Linear algebra shows that $(\lambda_1 q_1, \lambda_2 q_2, \lambda_3 q_3) = (f_4(\theta_1), f_4(\theta_2), f_4(\theta_3))$ for some f_4 if and only if $\lambda_1 b_0 = \lambda_2 c_4 = \lambda_3 d_0$, and $2\lambda_1 b_1 = \lambda_2 c_1 + \lambda_3 d_1$. There exist $\lambda_1, \lambda_2, \lambda_3 \neq 0$ such that $(\lambda_1 q_1, \lambda_2 q_2, \lambda_3 q_3) = (f_4(\theta_1), f_4(\theta_2), f_4(\theta_3))$ for some f_4 and such that $\mathbb{Z}(f_4)$ and $\mathbb{Z}(f_3)$ have no common singular point if and only if all of the following hold:

- $b_0 = 0 \leftrightarrow c_0 = 0 \leftrightarrow d_0 = 0$,
- if $b_0 = c_0 = d_0 = 0$, then at least two of b_1, c_1, and d_1 are different from zero,
- $2b_1 c_0 d_0 = b_0 c_1 d_0 + b_0 c_0 d_1$.

As in all the previous cases we can classify the possible configurations of other singularities for all possible j_1, j_2, j_3. As before, the first bullet point only affect the multiplicity of the factor t in q_1, q_2 and q_3. For any set of multiplicities for the rest of the roots, we can find q_1, q_2, q_3 with real roots of the given multiplicities such that the last bullet point is satisfied. This completes the classification of the singularities (other than O) when $\mathbb{Z}(f_3)$ is three lines meeting in a point.

Case 7. The tangent cone is a double line plus a line, and we can assume $f_3 = x_2 x_3^2$. The tangent cone is singular along the line $\mathbb{Z}(x_3)$. The line $\mathbb{Z}(x_2)$ is parameterized by θ_1 and the line $\mathbb{Z}(x_3)$ is parameterized by θ_2 where $\theta_1(s, t) = (s, 0, t)$ and $\theta_2(s, t) = (s, t, 0)$. The monoid is reducible if and only if $f_4(\theta_1)$ or $f_4(\theta_2)$ is identically zero, so we assume that neither is identically zero. For each f_4 we associate two integers, $j_0 := I_{(1:0:0)}(f_4, x_2)$ and $k_0 := I_{(1:0:0)}(f_4, x_3)$. Furthermore, we write $f_4(\theta_2)$ as a product of linear factors

$$f_4(\theta_2) = \lambda s^{k_0} \prod_{i=0}^{r} (\alpha_i s - t)^{m_i}.$$

Now the singularity at O will be of the V series and depends on j_0, k_0 and $m_1, \ldots, m_r$.

Other singularities on the monoid correspond to intersections of $\mathbb{Z}(f_4)$ and the line $\mathbb{Z}(x_2)$ away from $(1 : 0 : 0)$. Each such intersection corresponds to a root in the polynomial $f_4(\theta_1)$ different from $(1 : 0)$. Let $j_0' \leq 4$ and $k_0' \leq 4$ be integers such that $j_0 > 0 \leftrightarrow k_0 > 0$. Then, for any homogeneous polynomials q_1, q_2 in s, t of degree 4 such that s is a factor of multiplicity j_0' in q_1 and of multiplicity k_0' in q_2, there is a polynomial f_4 and nonzero constants λ_1 and λ_2 such that $k_0 = k_0'$, $j_0 = j_0'$ and $(\lambda_1 q_1, \lambda_2 q_2) = (f_4(\theta_1), f_4(\theta_2))$. Furthermore, if q_1 and q_2 have real coefficients, then f_4 can be selected with real coefficients. This follows from an analysis similar to case 5 and completes the classification of singularities when the tangent cone is a product of a line and a double line.

Case 8. The tangent cone is a triple line, and we assume that $f_3 = x_3^3$. The line $\mathbb{Z}(x_3)$ is parameterized by θ where $\theta(s, t) = (s, t, 0)$. Assume that the polynomial $f_4(\theta)$ has r distinct roots with multiplicities $m_1, \ldots, m_r$. (As before $f_4(\theta) \equiv 0$ if and only if the monoid is reducible.) Then the type of the singularity at O will be of the V' series [3, p. 267]. The integers $m_1, \ldots, m_r$ are constant under right equivalence over $\mathbb{C}$. Note that one can construct examples of monoids that are right equivalent over $\mathbb{C}$, but not over $\mathbb{R}$ (see Figure 4.4).

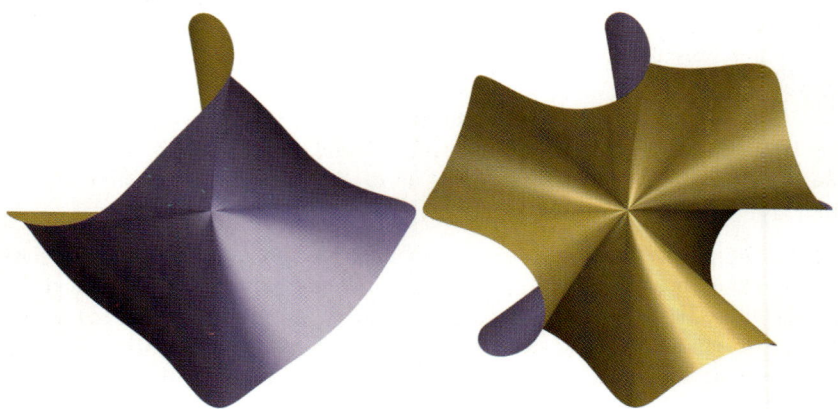

Fig. 4.4. The monoids $\mathbb{Z}(z^3 + xy^3 + x^3y)$ and $\mathbb{Z}(z^3 + xy^3 - x^3y)$ are right equivalent over $\mathbb{C}$ but not over $\mathbb{R}$.

The tangent cone is singular everywhere, so there can be no other singularities on the monoid.

Case 9. The tangent cone is a smooth cubic curve, and we write $f_3 = x_1^3 + x_2^3 + x_3^3 + 3ax_1x_2x_3$ where $a^3 \neq -1$. This is a one-parameter family of elliptic curves, so we cannot use the parameterization technique of the other cases. The singularity at O will be a P_8 singularity (cf. [3, p. 185]), and other singularities correspond to intersections between $\mathbb{Z}(f_3)$ and $\mathbb{Z}(f_4)$, as described by Proposition 6.

To classify the possible configurations of singularities on a monoid with a non-singular (projective) tangent cone, we need to answer the following question: For any positive integers $m_1, \ldots, m_r$ such that $\sum_{i=1}^{r} m_i = 12$, does there, for *some* $a \in \mathbb{R} \setminus \{-1\}$, exist a polynomial f_4 with real coefficients such that $\mathbb{Z}(f_3, f_4) = \{p_1, \ldots, p_r\} \in \mathbb{P}^2(\mathbb{R})$ and $I_{p_i}(f_3, f_4) = m_i$ for $i = 1, \ldots, r$? Rohn [15, p. 63] says that one can always find curves $\mathbb{Z}(f_3)$, $\mathbb{Z}(f_4)$ with this property. Here we shall show that for *any* $a \in \mathbb{R} \setminus \{-1\}$ we can find a suitable f_4.

In fact, in almost all cases f_4 can be constructed as a product of linear and quadratic terms in a simple way. The difficult cases are $(m_1, m_2) = (11, 1)$, $(m_1, m_2, m_3) = (8, 3, 1)$, and $(m_1, m_2) = (5, 7)$. For example, the case where $(m_1, m_2, m_3) = (3, 4, 5)$ can be constructed as follows: Let $f_4 = \ell_1 \ell_2 \ell_3^2$ where ℓ_1 and ℓ_2 define tangent lines at inflection points p_1 and p_3 of $\mathbb{Z}(f_3)$. Let ℓ_3 define a line that intersects $\mathbb{Z}(f_3)$ once at p_3 and twice at another point p_2. Note that the points p_1, p_2 and p_3 can be found for any $a \in \mathbb{R} \setminus \{-1\}$.

The case $(m_1, m_2) = (11, 1)$ is also possible for every $a \in \mathbb{R} \setminus \{-1\}$. For any point p on $\mathbb{Z}(f_3)$ there exists an f_4 such that $I_p(f_3, f_4) \geq 11$. For all except a finite number of points, we have equality [11], so the case $(m_1, m_2) = (11, 1)$ is possible for any $a \in \mathbb{R} \setminus \{-1\}$. The case $(m_1, m_2, m_3) = (8, 3, 1)$ is similar, but we need to let f_4 be a product of the tangent at an inflection point with another cubic.

The case $(m_1, m_2) = (5, 7)$ is harder. Let $a = 0$. Then we can construct a conic C that intersects $\mathbb{Z}(f_3)$ with multiplicity five in one point and multiplicity one in an inflection point, and choosing $\mathbb{Z}(f_4)$ as the union of C and twice the tangent line through the inflection point will give the desired example. The same can be done for $a = -4/3$. By using the computer algebra system SINGULAR [6] we can show that these constructions can be continuously extended to any $a \in \mathbb{R} \setminus \{-1\}$. This completes the classification of singularities on a monoid when the tangent cone is smooth.

In the Cases 3, 5, and 6, not all real equations of a given type can be transformed to the chosen forms by a real transformation.

In Case 3 the conic may not intersect the line in two real points, but rather in two complex conjugate points. Then we can assume $f_3 = x_3(x_1 x_3 + x_1^2 + x_2^2)$, and the singular points are $(1 : \pm i : 0)$. For any real f_4, we must have

$$I_{(1:i:0)}(x_1 x_3 + x_1^2 + x_2^2, f_4) = I_{(1:-i:0)}(x_1 x_3 + x_1^2 + x_2^2, f_4)$$

and

$$I_{(1:i:0)}(x_3, f_4) = I_{(1:-i:0)}(x_3, f_4),$$

so only the cases where $j_0 = j_1$ and $k_0 = k_1$ are possible. Apart from that, no other restrictions apply.

In Case 5, two of the lines can be complex conjugate, and we assume $f_3 = x_3(x_1^2 + x_2^2)$. A configuration from the previous analysis is possible for real coefficients of f_4 if and only if $m_1 = m_2$, $k_2 = l_1$, and $k_3 = l_3$. Furthermore, only the singularities that correspond to the line $\mathbb{Z}(x_3)$ will be real.

In Case 6, two of the lines can be complex conjugate, and then we may assume $f_3 = x_2^3 + x_3^3$. Now, if j_3 denotes the intersection number of $\mathbb{Z}(f_4)$ with the real line $\mathbb{Z}(x_2 + x_3)$, precisely the cases where $j_1 = j_2$ are possible, and only intersections with the line $\mathbb{Z}(x_2 + x_3)$ may contribute to real singularities.

This concludes the classification of real and complex singularities on real monoids of degree 4.

Remark. In order to describe the various monoid singularities, Rohn [15] computes the "class reduction" due to the presence of the singularity, in (almost) all cases. (The class is the degree of the dual surface [14, p. 262].) The class reduction is equal to the local intersection multiplicity of the surface with two general polar surfaces. This intersection multiplicity is equal to the sum of the Milnor number and the Milnor number of a general plane section through the singular point [20, Cor. 1.5, p. 320]. It is not hard to see that a general plane section has either a D_4 (Cases 1–6, 9), D_5 (Case 7), or E_6 (Case 8) singularity. Therefore one can retrieve the Milnor number of each monoid singularity from Rohn's work.

Acknowledgements

We would like to thank the referees for helpful comments. This research was supported by the European Union through the project IST 2001–35512 'Intersection algorithms for geometry based IT applications using approximate algebraic methods' (GAIA II).

References

1. V. I. Arnol'd. Normal forms of functions in neighbourhoods of degenerate critical points. *Russian Math Surveys*, 29ii:10–50, 1974.
2. V. I. Arnol'd. Critical points of smooth functions and their normal forms. *Russian Math Surveys*, 30v:1–75, 1975.
3. V. I. Arnol'd, S. M. Guseĭn-Zade, and A. N. Varchenko. *Singularities of differentiable maps. Vol. I*, volume 82 of *Monographs in Mathematics*. The classification of critical points, caustics and wave fronts, Translated from the Russian by Ian Porteous and Mark Reynolds. Birkhäuser Boston Inc., Boston, MA, 1985.
4. J. W. Bruce and C. T. C. Wall. On the classification of cubic surfaces. *J. London Math. Soc. (2)*, 19(2):245–256, 1979.
5. S. Endraß et al. SURF 1.0.4. 2003. A Computer Software for Visualising Real Algebraic Geometry, http://surf.sourceforge.net.
6. G.-M. Greuel, G. Pfister, and H. Schönemann. SINGULAR 3.01. A Computer Algebra System for Polynomial Computations. Centre for Computer Algebra, University of Kaiserslautern (2005). http://www.singular.uni-kl.de.
7. C. M. Jessop. Quartic surfaces with singular points. *Cambridge University Press*, 1916.
8. H. Knörrer and T. Miller. Topologische Typen reeller kubischer Flächen. *Math. Z.*, 195(1):51–67, 1987.
9. O. Labs and D. van Straten. The Cubic Surface Homepage. 2001. http://Cubics.AlgebraicSurface.net.
10. H. B. Laufer. On minimally elliptic singularities. *Amer. J. Math.*, 99(6):1257–1295, 1977.
11. C. Moura. Local intersections of plane algebraic curves. *Proc. Amer. Math. Soc.*, 132(3):687–690 (electronic), 2004.
12. S. Pérez-Díaz, J. Sendra, and J. Rafael Sendra. Parametrization of approximate algebraic curves by lines. *Theoret. Comput. Sci.*, 315(2-3):627–650, 2004.
13. S. Pérez-Díaz, J. Sendra, and J. R. Sendra. Parametrization of approximate algebraic surfaces by lines. *Comput. Aided Geom. Design*, 22(2):147–181, 2005.
14. R. Piene. Polar classes of singular varieties. *Ann. Sci. École Norm. Sup. (4)*, 11(2):247–276, 1978.
15. K. Rohn. Ueber die Flächen vierter Ordnung mit dreifachem Punkte. *Math. Ann.*, 24(1):55–151, 1884.
16. L. Schläfli. On the distribution of surfaces of the third order into species. *Philos. Trans. Royal Soc.*, CLIII:193–241, 1863.
17. T. W. Sederberg, J. Zheng, K. Klimaszewski, and T. Dokken. Approximate implicitization using monoid curves and surfaces. *Graphical Models and Image Processing*, 61(4):177–198, 1999.
18. B. Segre. *The Non-singular Cubic Surfaces*. Oxford University Press, Oxford, 1942.
19. T. Takahashi, K. Watanabe, and T. Higuchi. On the classification of quartic surfaces with triple point. I, II. *Sci. Rep. Yokohama Nat. Univ. Sect. I*, (29):47–70, 71–94, 1982.

20. B. Teissier. Cycles évanescents, sections planes et conditions de Whitney. In *Singularités à Cargèse (Rencontre Singularités Géom. Anal., Inst. Études Sci., Cargèse, 1972)*, pages 285–362. Astérisque, Nos. 7 et 8. Soc. Math. France, Paris, 1973.
21. Y. Umezu. On normal projective surfaces with trivial dualizing sheaf. *Tokyo J. Math.*, 4(2):343–354, 1981.

5

Canal Surfaces Defined by Quadratic Families of Spheres

Rimvydas Krasauskas and Severinas Zube

Faculty of Mathematics and Informatics,
Vilnius University
rimvydas.krasauskas@maf.vu.lt

Summary. This paper is devoted to quadratic canal surfaces, i.e. surfaces that are envelopes of quadratic families of spheres. They are generalizations of Dupin cyclides but are more flexible as blending surfaces between natural quadrics . The classification of quadratic canal surfaces is given from the point of view of Laguerre geometry. Their properties that are important for geometric modeling are studied: rational parametrizations of minimal degree, Bézier representations, and implicit equations.

5.1 Introduction

Natural quadrics, i.e. spheres, circular cylinders and circular cones, are perhaps the most popular surfaces in geometric modeling. They can be characterized as envelopes of linear (or constant) families of spheres in space. An other exceptional property of natural quadrics is that their offset surfaces are of the same type. Usually Dupin cyclides are used as blending surfaces between natural quadrics. For example, any two circular cones with a common inscribed sphere can be blended by a part of Dupin cyclide bounded by two circles as it was shown by Pratt [9] (see Fig. 5.1). Cyclides are envelopes of special quadratic families of spheres, and they are offset stable as well. Here we consider envelopes of most general quadratic families of spheres and call them *quadratic canal surfaces*. The main motivation is the possibility to use patches of these surfaces for blending of natural quadrics.

In Section 5.2 we briefly remaind elements of Laguerre geometry. Section 5.3 is devoted to the classification of conics in the Laguerre space. Cases when conics define quadratic canal surfaces that can be tangent to natural cones along non-trivial curves are determined. In Section 5.4 we find rational parametrizations of such canal surfaces of minimal degree. Their Bézier representations and implicit equations are considered in Sections 5.5 and 5.6. Conclusions and possible applications are discussed Section 5.7.

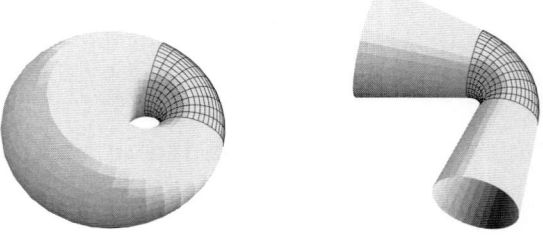

Fig. 5.1. A Dupin cyclide used for blending circular cones.

5.2 Elements of Laguerre geometry

Here we briefly recall the elements of Laguerre Geometry (cf. [5, 8]). Consider the space of all oriented spheres in $\mathbb{R}^3$ as a 4-dimensional affine space $\mathbb{R}^4$, where the first three coordinates (x_1, x_2, x_3) are the center point of a sphere and the last coordinate x_4 represents the radius of the sphere. The orientation is defined by the sign of x_4: the normals are pointing outwards if $x_4 > 0$. We denote by $\mathbb{R}_1^4$ the affine space $\mathbb{R}^4$ equipped with a *pseudo-euclidean* (PE) metrics defined by the following *PE scalar product* of vectors v, v':

$$\langle v, v' \rangle = v_1 v_1' + v_2 v_2' + v_3 v_3' - v_4 v_4'. \tag{5.1}$$

Affine transformations of $\mathbb{R}_1^4$ that preserve this PE scalar product are called *PE transformations*. It will be convenient to consider also the projective extension $\mathbb{R}P^4$ of $\mathbb{R}P_1^4$ with additional coordinate x_0. From this projective point of view PE transformations are exactly those projective transformations of $\mathbb{R}P^4$ that preserve the *absolute quadric* Ω: $x_0 = 0$, $x_1^2 + x_2^2 + x_3^2 - x_4^2$. A geometric meaning of this metric is a tangential distance between spheres in $\mathbb{R}^3$.

Affine subspaces $A \subset \mathbb{R}_1^4$ can be of three signature types sign $A = (+, \ldots, +, \sigma)$, where $\sigma \in \{+, 0, -\}$. For example, all lines in $\mathbb{R}_1^4$ with directional vectors v can be classified into three types depending on the sign $\sigma = \text{sign}\langle v, v \rangle$: $(+)$-lines, (0)-lines, and $(-)$-lines (also called *positive, isotropic*, and *negative* lines, resp.). 2-Planes also can have three types: $(++)$-, $(+0)$-, $(+-)$-planes.

For any smooth curve $\alpha \subset \mathbb{R}_1^4$ with tangent $(+)$-lines almost everywhere, define $\text{Env}(\alpha)$ as an envelope of the corresponding family of spheres in $\mathbb{R}^3$. We call such envelopes also canal surfaces. Circular cylinder or circular cones (call them both *natural cones*) are envelopes $\text{Env}(L)$ of $(+)$-lines L and vice versa. Let $\Gamma(a)$ denote the hypersurface $\langle x - a, x - a \rangle = 0$. Define Γ-hypersurface $\Gamma(\alpha)$ of a smooth curve α in $\mathbb{R}^4$ as the envelope of the family of $\Gamma(\alpha(t))$, for all t. Then $\text{Env}(\alpha) = \Gamma(\alpha) \cap \{x_4 = 0\}$ and any point $x \in \Gamma(\alpha)$ corresponds to a sphere that touches the canal surface $\text{Env}(\alpha)$. Therefore, for any other curve β canal surfaces $\text{Env}(\alpha)$ and $\text{Env}(\beta)$ touch each other along some curve if and only if $\beta(t) \in \Gamma(\alpha)$ for all t in some open interval. Here we exclud the trivial case when α and β are tangent in a common point (hence the canal surfaces touch along a circle).

Let C be a natural cone, and let L be the corresponding hyperbolic line in $\mathbb{R}^4$. All the spheres in $\mathbb{R}^3$ tangent to C correspond to an *isotropic hypersurface* $\Gamma(L)$ in $\mathbb{R}_1^4$. Let v be a directional vector of L, and let p be a point on L. The equation of $\Gamma(L)$ can be calculated easily (see [2])

$$\Gamma(L): \quad \langle x - p, x - p \rangle \langle v, v \rangle - \langle x - p, v \rangle^2 = 0. \tag{5.2}$$

5.3 Conics in $\mathbb{R}_1^4$

Our goal is to study quadratic canal surfaces . Since they are defined as envelopes of quadratic families of spheres, they are encoded by conics in $\mathbb{R}_1^4$. The well-known examples are Dupin cyclides (see [5, 9]). The corresponding conics C are characterized as PE circles, i.e. infinite points of C are lying on Ω (may be a pair of complex conjugated points or a double point). For example, all conics C contained in $\Gamma(a)$, $a \in \mathbb{R}^4$ are PE circles. Therefore, $\mathrm{Env}(C)$ is a Dupin cyclide [5, 8].

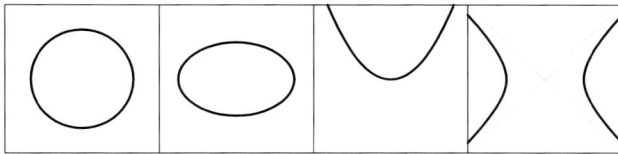

Fig. 5.2. Conics of type $\sigma = (++)$ in canonical position.

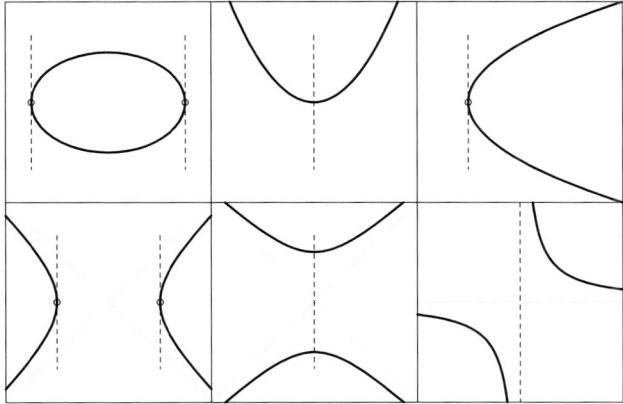

Fig. 5.3. Conics of type $\sigma = (+0)$ in canonical position.

Let us classify PE types of conics in $\mathbb{R}_1^4$ by:

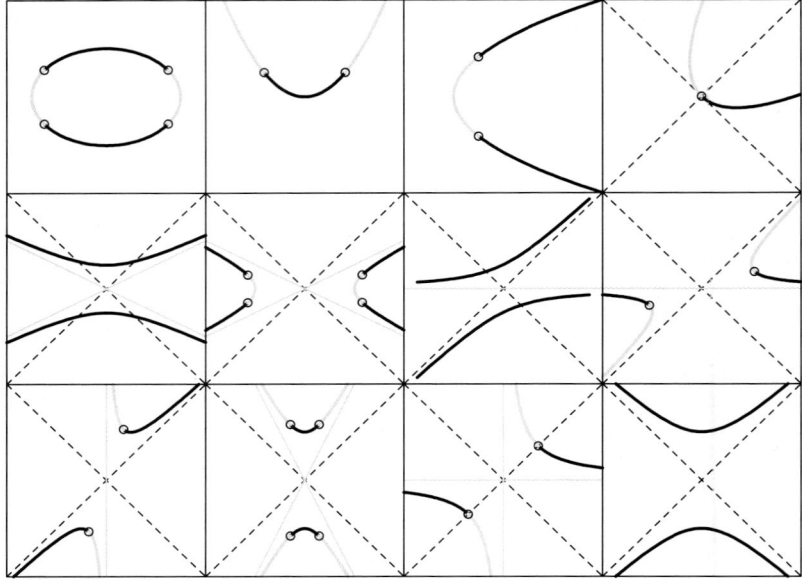

Fig. 5.4. Conics of type $\sigma = (+-)$ in canonical position.

- an affine type $\mathcal{T} = \mathcal{E}, \mathcal{P}, \mathcal{H}$: ellipse, parabola, or hyperbola;
- a signature $\sigma = (++), (+0), (+-)$ of the spanned affine 2-plane;
- positions of infinite points of C with respect to Ω, e.g. a number $n = \#(C \cap \Omega)$.

We denote the class $\mathcal{T}_\sigma^{[n]}$, where we skip $[n]$ if $n = 0$, or change $[n]$ to the list of signatures $\alpha = (\alpha_1, \alpha_2)$ of asymptotic directions when it is necessary. In hyperbolic case $\mathcal{H}$ sometimes it is necessary to distinguish conjugated hyperbolas having the same asymptotes. We mark with a tilde $\widetilde{\mathcal{H}}_\sigma^\alpha$ a case that is conjugated to $\mathcal{H}_\sigma^\alpha$.

Theorem 1. *All PE equivalence classes of irreducible conics (i.e. not a pair of lines) with non-empty set of real points in $\mathbb{R}_1^4$ are listed in the following table, except three conjugated cases with totally negative tangents $\widetilde{\mathcal{H}}_{+-}^{[2]}, \widetilde{\mathcal{H}}_{+-}^{(0-)}, \widetilde{\mathcal{H}}_{+-}^{(--)}$:*

$\sigma = ++$	$\mathcal{E}_{++}^{[2]}$	$\mathcal{E}_{++}$	$\mathcal{P}_{++}$	$\mathcal{H}_{++}$
$\sigma = +0$	$\mathcal{E}_{+0}$	$\mathcal{P}_{+0}^{[2]}$	$\mathcal{P}_{+0}$	
	$\widetilde{\mathcal{H}}_{+0}$	$\mathcal{H}_{+0}$	$\mathcal{H}_{+0}^{[1]}$	
$\sigma = +-$	$\mathcal{E}_{+-}$	$\mathcal{P}_{+-}^{(-)}$	$\mathcal{P}_{+-}^{(+)}$	$\mathcal{P}_{+-}^{(0)}$
	$\mathcal{H}_{+-}^{(++)}$	$\widetilde{\mathcal{H}}_{+-}^{(++)}$	$\mathcal{H}_{+-}^{(+0)}$	$\widetilde{\mathcal{H}}_{+-}^{(+0)}$
	$\mathcal{H}_{+-}^{(0-)}$	$\mathcal{H}_{+-}^{(--)}$	$\mathcal{H}_{+-}^{(+-)}$	$\mathcal{H}_{+-}^{[2]}$

Conics with different signatures σ from this table are illustrated in Fig. 5.2–5.4. Arcs of curves with negative tangents are shown in grey. Points with isotropic *tangents are marked by small circles. Isotropic directions and asymptotes are shown by thin dashed and grey lines, respectively.*

Proof. Without loss of generality we suppose that a 2-plane containing the given conic passes through the origin and has the basis $\{e_1, e_2\}$, $\{e_1, e_3 + e_4\}$, $\{e_3, e_4\}$, depending on the signature $\sigma = (++), (+0), (+-)$, respectively. Also let the center of the conic (or the vertex in parabolic case) be in the origin. Then in all cases there exist linear PE transformations that have the following matrix form (when restricted to these 2-planes with the fixed basis):

$$\begin{pmatrix} \cos\varphi & \mp\sin\varphi \\ \sin\varphi & \pm\cos\varphi \end{pmatrix}, \quad \begin{pmatrix} 1 & 0 \\ \rho & \pm 1 \end{pmatrix}, \quad \pm\begin{pmatrix} \cosh\theta & \pm\sinh\theta \\ \sinh\theta & \pm\cosh\theta \end{pmatrix}, \quad \varphi, \rho, \theta \in \mathbb{R}. \quad (5.3)$$

It is easy to recognize rotation, shear and boost (or hyperbolic rotation) transformations possibly composed with reflections.

   Case $\sigma = (++)$. $\mathcal{E}_{++}$, $\mathcal{P}_{++}$, $\mathcal{H}_{++}$ are usual Euclidean types of ellipse, parabola and hyperbola that can be rotated to the canonical positions. Here we distinguish a circle case $\mathcal{E}_{++}^{[2]}$, since it has two 'circular points' $(0, 1, \pm i, 0, 0)$ lying on Ω.

   Case $\sigma = (+0)$. Only one direction is isotropic (shown as dashed vertical lines in Fig. 5.3) and all others are positive. Any positive direction can be moved to any other positive one using a shear transformation. Hence an axis of a parabola $\mathcal{P}_{+0}$ and an asymptote of a hyperbola $\mathcal{H}_{+0}^{[1]}$ can be moved to the horizontal position. Similarly asymptotes of a hyperbola $\mathcal{H}_{+0}$ can be transformed to the symmetric position. The conjugated hyperbola $\widetilde{\mathcal{H}}_{+0}$ has a different PE type, since it has isotropic tangents. An ellipse $\mathcal{E}_{+0}$ has a pair of points with isotropic tangents, and one can move line connecting these points to the horizontal position.

   Case $\sigma = (+-)$. There are two fixed isotropic directions. Positive and negative directions are in between. Using boost transformation one can move any positive (resp. negative) direction to the horizontal (resp. vertical) direction. This allows to transform all cases to the canonical ones shown in Fig. 5.4. For example, in the case $\mathcal{E}_{+-}$ we choose a vector connecting the origin with a point on the ellipse with a biggest PE length, and transform it to a horizontal position using an appropriate boost.

Corollary 2. *Let L be a $(+)$-line, and C is a conic in $\mathbb{R}^4$. If the cone $\mathrm{Env}(L)$ touches the quadratic canal surface $nv(C)$ along a curve which is neither a line nor a circle then C has one of the following types: $\mathcal{E}_{++}$, $\mathcal{P}_{++}$, $\mathcal{H}_{++}$, $\mathcal{P}_{+0}^{[2]}$, $\mathcal{H}_{+0}$, $\mathcal{H}_{+-}^{(++)}$, $\mathcal{H}_{+-}^{(+0)}$.*

Proof. Without loss of generality we identify L with the x_1-axis. Then by Eq. (5.2) $\Gamma(L)$ has the equation $x_2^2 + x_3^2 = x_4^2$. Consider a projection $\pi : (x_1, x_2, x_3, x_4) \mapsto (0, x_2, x_3, x_4)$ to the hyperplane $\{x_1 = 0\} \subset \mathbb{R}^4$. The conic C is contained in $\Gamma(L)$ (since the touching curve is non circular), and its projection $\pi(C)$ is a conic in $\pi(\Gamma(L))$ (since the touching curve is not a line). On the other hand $\pi(\Gamma(L)) = \Gamma(L) \cap \{x_1 = 0\}$, and all its infinite points are contained in the absolute quadric Ω. Hence $\pi(C)$ has infinite points on Ω, so it is one of three PE circles $\mathcal{E}_{++}^{[2]}$, $\mathcal{P}_{+0}^{[2]}$, $\mathcal{H}_{+-}^{[2]}$. Note that Ω in the infinite hyperplane $x_0 = 0$ has the same equation as a sphere, where x_4 plays a role of x_0. Let $\lambda(C)$ be an infinite line of the 2-plane spanned by the conic C. Different signatures of the 2-plane $\sigma = (++), (+0), (+-)$ correspond to different number of intersection points $\#(\lambda(C) \cap \Omega) = 0, 1, 2$, respectively. The proof now follows from the following simple observations:

(i) the projection π preserves affine type of conics;
(ii) the tangent line to the conic C in any point must be positive, since tangents to $\pi(C)$ are positive.

Indeed, according to (ii) all cases can be directly chosen from Fig. 2-4 with one exception: $\mathcal{H}_{+0}^{[1]}$ cannot be included in the list, since $\lambda(C)$ is tangent to Ω in the point that already belongs to C. This means that $\pi(C)$ should be a parabola $\mathcal{P}_{+0}^{[2]}$, that is impossible according to (i).

5.4 Rational parametrizations

Any rational curve $C \subset \mathbb{R}_1^4$ with non-negative tangents defines a canal surface $\mathrm{Env}(C)$ with a rational spine curve $s(t) = (C_1(t), C_2(t), C_3(t))$ and a rational radius function $r(t) = C_4(t)$. It is known that there is a rational parametrization of such canal surface in the form

$$M(t, u) = s(t) + r(t)N(t, u), \tag{5.4}$$

where $N(t, u)$ defines a rational Gaussian map to the unit sphere S. Let $c(t)$ and $n(t, u)$ be a homogeneous form of $C(t)$ and $N(t, u)$, i.e. $C(t) = (c_1/c_0, \dots, c_4/c_0)$ and $N = (n_1/n_0, n_2/n_0, n_3/n_0)$. In Laguerre geometry it is natural to consider a slightly different variant of a Gaussian map with the image at infinity $\tilde{n} = (0, n_1, n_2, n_3, -n_0)$. Note that $\tilde{n}(t, u) \in \Omega$.

Lemma 3. *An isotropic hypersurface $\Gamma(C) \subset \mathbb{R}_1^4$ can be parametrized by 2-parameter set of isotropic lines connecting $c(t)$ with $\tilde{n}(t, u)$ for all $t, u \in \mathbb{R}P^1$.*

Proof. Any of such lines can be parametrized $v_0 c(t) + v_1 \tilde{n}(t, u)$ with homogeneous coordinates $(v_0 : v_1)$. The intersection with the hyperplane $x_4 = 0$ gives the condition $v_0 c_4(t) - v_1 n_0(t, u) = 0$. Hence choosing $v_0 = n_0(t, u)$ and $v_1 = c_4(t)$ we get the parametrization of the intersection $\Gamma(C) \cap \{x_4 = 0\}$:

$$n_0(t, u)c(t) + c_4(t)\tilde{n}(t, u) = (n_0 c_0, n_0 c_1 + c_4 n_1, n_0 c_2 + c_4 n_2, n_0 c_3 + c_4 n_3, 0).$$

Switching to the cartesian coordinates in $\mathbb{R}^3$ we get exactly the parametrization of the canal surface $\mathrm{Env}(C)$ (5.4).

Using this lemma one can easily find a rational parametrization of any PE transform of the canal surface. It is enough to transform the curve $c(t)$ and the Gaussian map $\tilde{n}(t, u)$ separately, and then intersect the resulting isotropic hypersurface with the hyperplane $x_4 = 0$. Hence it remains to find the Gaussian map in all canonical cases of interest. Here we remind some definitions and results from [3].

For a rational curve $C(t) \subset \mathbb{R}_1^4$ define its *discriminant* by the formula $\mathcal{D}(t) = d^2(t)(\dot{C}_1^2(t) + \dot{C}_2^2(t) + \dot{C}_3^2(t) - \dot{C}_4^2(t))$, where $d(t)$ is a common denominator of the derivative vector $\dot{C}(t)$. Let S be a unit sphere in $\mathbb{R}P^3$. Define the parametrization $\mathcal{P}_S$ of S by two complex parameters u_0 and u_1

$$\mathcal{P}_S(u_0, u_1) = (u_0\bar{u}_0 + u_1\bar{u}_1, 2\mathrm{Re}(u_0\bar{u}_1), 2\mathrm{Im}(u_0\bar{u}_1), u_0\bar{u}_0 - u_1\bar{u}_1). \qquad (5.5)$$

For any 2×2-matrix $X = (x_{ij})$ with complex entries define the following extended 2×4-matrix $\widetilde{X}$ and its minors $q_{ij} = q_{ij}(\widetilde{X})$:

$$\widetilde{X} = \begin{pmatrix} x_{00} & x_{01} & \bar{x}_{00} & \bar{x}_{01} \\ x_{10} & x_{11} & \bar{x}_{10} & \bar{x}_{11} \end{pmatrix}, \quad \text{i.e. } q_{01} = \det X, \quad q_{02} = \begin{vmatrix} x_{00} & \bar{x}_{00} \\ x_{10} & \bar{x}_{10} \end{vmatrix}, \quad \text{etc. } (5.6)$$

Theorem 4. *For any given curve $C \subset \mathbb{R}_1^4$ with $D(t) \geq 0$ the Gaussian map of the canal surface $\mathrm{Env}(C)$ has the form*

$$n(t, u) = \mathcal{P}_S(x_{1*}(t)(1 - u) + x_{2*}(t)u), \qquad (5.7)$$

where x_{i*} are rows of the 2×2 complex polynomial matrix $X(t)$ such that the minors $q_{ij} = q_{ij}(\widetilde{X}(t))$ (see (5.6)) satisfy the following condition

$$2\mathrm{Im}(q_{12}) : 2\mathrm{Re}(q_{12}) : \mathrm{Im}(q_{13} - q_{02}) : \mathrm{Im}(q_{13} + q_{02}) = \dot{C}_1 : \dot{C}_2 : \dot{C}_3 : \dot{C}_4. \quad (5.8)$$

If $n(t, u)$ has minimal degree in t then $\deg x_{0*}(t) + \deg x_{1*}(t) = \deg(d(t)\dot{C}(t))$. All such cases correspond to different factorizations of the discriminant $D(t) = q_{01}\bar{q}_{01}$.

Proof. The proof follows directly from [3, Sec. 4]. In particular the condition (5.8) follows from the following equation (cf. [3, Eq. (7)])

$$(q_{13}, -q_{12}, -q_{03}, q_{02}) = d\mathrm{i}(\dot{r} + \dot{s}_3, \dot{s}_1 - \mathrm{i}\dot{s}_2, \dot{s}_1 + \mathrm{i}\dot{s}_2, \dot{r} - \dot{s}_3).$$

Now we are ready to consider minimal parametrizations of six cases of quadratic canal surfaces $\mathcal{E}_{++}$, $\mathcal{P}_{++}$, $\mathcal{H}_{++}$, $\mathcal{H}_{+0}$, $\mathcal{H}_{+-}^{(++)}$, $\mathcal{H}_{+-}^{(+0)}$. We skip parabolic Dupin cyclide case $\mathcal{P}_{+0}^{[2]}$, since it was already considered earlier ([1, 5, 8]).

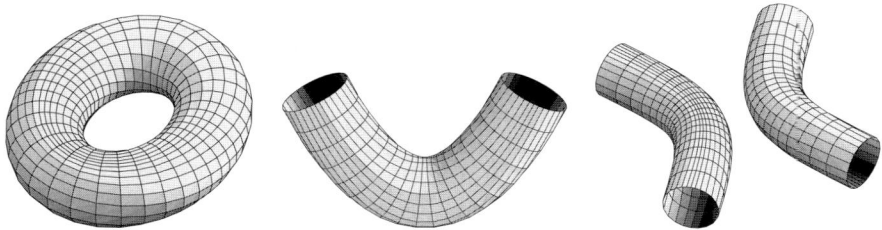

Fig. 5.5. Quadratic canal surfaces of types $\mathcal{E}_{++}$, $\mathcal{P}_{++}$, $\mathcal{H}_{++}$.

Example 5. The case $\mathcal{E}_{++}$ is defined by the ellipse $c(t) = (1+t^2, a(1-t^2), 2bt, 0, 0)$, $a = p^2 + q^2$, $b = 2pq$, $p > q > 0$. Then $D(t) = (p^2 t^2 + q^2)(q^2 t^2 + p^2)$ has two pairs of complex conjugated roots and can be factorized $D(t) = f(t)\bar{f}(t)$ in four different ways corresponding to four complex matrices $X(t)$

$$\begin{pmatrix} b(t^2 - 1) - 2iat & 0 \\ 0 & 1 \end{pmatrix}, \begin{pmatrix} 1 & 0 \\ 0 & b(t^2 - 1) + 2iat \end{pmatrix},$$

$$\begin{pmatrix} pt - iq & 0 \\ 0 & qt + ip \end{pmatrix}, \begin{pmatrix} qt - ip & 0 \\ 0 & pt + iq \end{pmatrix},$$

such that $\det X(t) = f(t)$. One can check straightforward that the minors q_{01} satisfy (5.8). Hence, the first two cases of $X(t)$ define Gaussian maps $n(t, u)$ (see (5.7)) of bi-degree $(4, 2)$ and the last two ones are biquadratic. Therefore, only the latter two cases define two different parametrizations $F(t, u)$ of $\mathrm{Env}(C)$ of bi-degree $(4, 2)$ (see (5.4)). One of such parametrizations is shown in Fig. 5.5(left). The other parametrization can be obtained by reflection in the plane $x = 0$.

Similar approach allows us to find all parametrizations of minimal degree for other quadratic canal surfaces we are considering.

Example 6. The case $\mathcal{P}_{++}$ is defined by the curve $c(t) = (1, 2a, at^2, 0, 0)$, $a > 0$. Then $D(t) = a^2(1 + t^2)$, and there are two bi-degree $(4, 2)$ parametrizations of $\mathrm{Env}(C)$ (Fig. 5.5(middle)) defined by the following matrices:

$$X(t) = \begin{pmatrix} a(t - i) & 0 \\ 0 & 1 \end{pmatrix}, \begin{pmatrix} 1 & 0 \\ 0 & a(t + i) \end{pmatrix}.$$

Example 7. The case $\mathcal{H}_{++}$: $c(t) = (1 - t^2, a(1 + t^2), 2bt, 0, 0)$, $a = p^2 - q^2$, $b = 2pq$, $p, q > 0$. Then $D(t) = (p^2 t^2 + q^2)(q^2 t^2 + p^2)$, and there are two bi-degree $(4, 2)$ parametrizations of $\mathrm{Env}(C)$ (Fig. 5.5(right)) defined by the following matrices:

$$X(t) = \begin{pmatrix} pt + iq & 0 \\ 0 & qt + ip \end{pmatrix}, \begin{pmatrix} qt - ip & 0 \\ 0 & pt - iq \end{pmatrix}.$$

Example 8. The case $\mathcal{H}_{+-}^{(++)}$: $c(t) = (1 - t^2, 0, 0, 2at, b(1 + t^2))$, $a = p^2 + q^2$, $b = 2pq$, $p, q > 0$. Then $D(t) = 1 + 2(1 - 2b^2/a^2)t^2 + t^4$, and there are four different bi-degree $(4, 2)$ parametrizations of $\mathrm{Env}(C)$ defined by the following matrices $X(t)$:

$$\begin{pmatrix} \bar{\lambda}t - 1 & t + \lambda \\ t - \lambda & \bar{\lambda}t + 1 \end{pmatrix}, \begin{pmatrix} \lambda t - 1 & t + \bar{\lambda} \\ t - \bar{\lambda} & \lambda t + 1 \end{pmatrix}, \begin{pmatrix} i(\lambda t - 1) & \bar{\lambda}t + 1 \\ i(t - \bar{\lambda}) & t + \lambda \end{pmatrix}, \begin{pmatrix} i(\bar{\lambda}t - 1) & \lambda t + 1 \\ i(t - \lambda) & t + \bar{\lambda} \end{pmatrix},$$

where $\lambda = -(q + ip)/(p + iq)$. In fact $\mathrm{Env}(C)$ is a hyperboloid of revolution (Fig. 5.6, left and right), or its offsets if C is translated in the x_4-axis direction (as shown in Fig. 5.6(middle)).

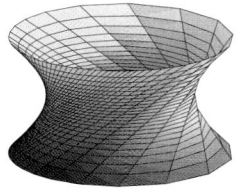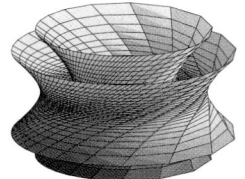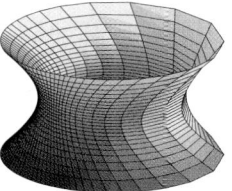

Fig. 5.6. Quadratic canal surfaces of type $\mathcal{H}_{+-}^{(++)}$.

Example 9. The case $\mathcal{H}_{+0}$: $c(t) = (1 - t^2, 2at, 0, b(1 + t^2), b(1 + t^2))$, $a > 0$. Then $D(t) = 4a^2b^2(1 + t^2)^2$, and there are two bi-degree $(4, 2)$ parametrizations of $\mathrm{Env}(C)$ (Fig. 5.7(left)) defined by the following matrices $X(t)$:

$$\begin{pmatrix} 2ib(t + i) & 0 \\ 2b & a(t + i) \end{pmatrix}, \quad \begin{pmatrix} 2ib(t - i) & 0 \\ 2b & a(t - i) \end{pmatrix}.$$

Example 10. The case $\mathcal{H}_{+-}^{[1]}$: $c(t) = (1 - t^2, 0, 0, 2at, a(1 + t)^2/2)$. Then $D(t) = a^2(3 + 2t + 3t^2)(1 - t)^2$, and there are two bi-degree $(3, 2)$ parametrizations of $\mathrm{Env}(C)$ (Fig. 5.7(right)) defined by the following matrices $X(t)$:

$$\begin{pmatrix} \sqrt{3}a(t + \mu) & -ia(t - 1) \\ -i\sqrt{3}(t + \mu) & (t - 1) \end{pmatrix}, \quad \begin{pmatrix} \sqrt{3}a(t + \bar{\mu}) & -ia(t - 1) \\ -i\sqrt{3}(t + \bar{\mu}) & (t - 1) \end{pmatrix}, \quad \mu = \frac{1 + 2\sqrt{2}i}{3}.$$

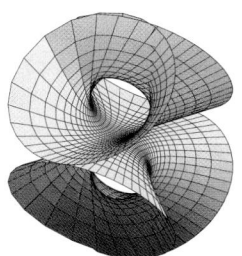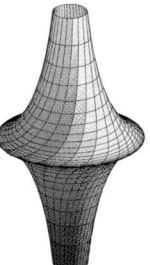

Fig. 5.7. Quadratic canal surfaces of type $\mathcal{H}_{+0}^{(++)}$ and $\mathcal{H}_{+-}^{[1]}$.

5.5 Bézier representations

Consider the case $\mathcal{E}_{++}$ (see Fig. 5.5(left)) of canal surface generated by the ellipse C: $x_1^2/a^2 + x_2^2/b^2 = x_0^2$ on the 2-plane $x_3 = x_4 = 0$. In order to parameterize $\mathrm{Env}(C)$ and its PE transforms we find a rational Bézier representation of $\Gamma(C)$ first.

We use homogeneous control points, denoting by $e_0, \ldots, e_4$ the standard frame in $\mathbb{R}P^4$. Consider the usual parametrization of a half of C: $c(t) = \sum_{i=0}^{2} p_i B_i^2(t)$ with control points

$$(p_0, p_1, p_2) = (e_0 - be_2, ae_1, e_0 + be_2).$$

Define a biquadratic parametrization $h(t, u) = \sum_{i=0}^{2} \sum_{j=0}^{2} r_{ij} B_i^2(t) B_j^2(u)$, where

$$(r_{ij}) = \begin{pmatrix} ab(e_3 + e_4) & ae_0 & ab(-e_3 + e_4) \\ -b\sqrt{a^2 - b^2}(e_3 + e_4) & (a^2 - b^2)e_1 & b\sqrt{a^2 - b^2}(-e_3 + e_4) \\ ab(e_3 + e_4) & ae_0 & ab(-e_3 + e_4) \end{pmatrix}. \quad (5.9)$$

This parametrization is derived from Example 5: take the parametrization $F(t, u)$ of $\Gamma(C)$ and intersect with the hyperplane $x_2 = 0$.

Then another rational parametrization of $\Gamma(C)$ can be defined by drawing lines through the ellipse points $c(t)$ and the hyperboloid points $h(t, u)$. The parametrization of the intersection of $\Gamma(C)$ with a hyperplane $\Pi(x) = 0$

$$f(t, u) = \Pi(h(t, u))c(t) - \Pi(c(t))h(t, u). \quad (5.10)$$

Now it is easy to calculate Bézier control points of this parametrization of bi-degree $(4, 2)$. Indeed, just use conversion formulas from products $B_i^2(t)B_j^2(t)$ to $B_{i+j}^4(t)$.

If we choose a hyperplane $\Pi(x) = x_4 - rx_0$ then the section is a pipe surface shown in Fig. 5.5(left).

5.6 Implicit equations and double points

It will be convenient to use affine coordinates $X = (x, y, z, r) = (x_1/x_0, \ldots, x_4/x_0)$ in $\mathbb{R}_1^4$. We start from the case $\mathcal{E}_{++}$ of ellipse. By the definition $\Gamma(E)$ is the envelope of the family of isotropic cones $\Gamma(E(t)) = \langle X - E(t), X - E(t) \rangle$, where $E(t) = (a(1 - t^2)/(1 + t^2), 2bt/(1 + t^2), 0, 0)$. The equation of the envelope is obtained by elimination of parameter t from the system

$$\begin{cases} f_1(X, t) = \langle X - E(t), X - E(t) \rangle = 0, \\ f_2(X, t) = \langle \dot{E}(t), X - E(t) \rangle = 0. \end{cases} \quad (5.11)$$

Here f_1, f_2 are rational functions in the variable t, i.e. $f_1 = g_1(X, t)/p(t)$, $f_2 = g_2(X, t)/q(t)$. Let

$$F(X) = \mathrm{Res}\left(g_1(X, t), g_2(X, t), t\right) \quad (5.12)$$

be the resultant of two polynomials g_1, g_2 (here we assume that degree of polynomials p, q are minimal). A priori F may be a reducible polynomial, i.e. $F = F_1 F_2 \cdots F_n$ then one factor, assume F_1, is the equation of $\Gamma(E)$. After easy computation with MAPLE package we see that degree of F is 8.

For example, if $a = \sqrt{2}$ and $b = 1$ the equation of $\Gamma(E)$ is

$$F = \omega^4 + \left(6 + 2x^2 - 4\,y^2\right)\omega^3 + \left(13 - 28\,y^2 + 4\,x^2 + \left(x^2 + 2\,y^2\right)^2\right)\omega^2$$
$$+\, 6\left(x^2 + 2 - 4\,y^2\right)\left(-x^2 + 1 - 2\,y^2\right)\omega$$
$$+\, \left(4 - 12\left(x^2 + 2\,y^2\right) - 3\left(x^2 - 4\,y^2\right)\left(4\,y^2 + 5\,x^2\right)\right) - 4\left(x^2 + 2\,y^2\right)^3,$$

where $\omega = x^2 + y^2 + z^2 - r^2$.

Moreover, F is an irreducible polynomial. Indeed, a plane curve $\Gamma(E) \cap \{r = const\} \cap \{z = 0\}$ is defined by an irreducible polynomial of degree 8 as it is an offset of the ellipse E (see, e.g. [8]). Hence $\deg \Gamma(E) = 8$ as well.

Theorem 11. *Let E be an ellipse of type $\mathcal{E}_{++}$. Then the hypersurface $\Gamma(E)$ is a real 3-dimensional variety of degree 8, and its set of finite real double points has four parts $E \cup H_1 \cup H_2 \cup \Omega$ where:*

$$E: \qquad x^2/a^2 + y^2/b^2 = 1, \qquad\qquad z = r = 0,$$
$$H_1: \quad x^2/(a^2 - b^2) - z^2/b^2 + r^2/b^2 = 1, \qquad y = 0,$$
$$H_2: -y^2/(a^2 - b^2) - z^2/a^2 + r^2/a^2 = 1, \qquad x = 0,$$

H_1 and H_2 are hyperboloids of 1-sheet and 2-sheet, resp., (the only non-isotropic lines on $\Gamma(E)$ are two rulings of the hyperboloid H_1), Ω is the absolute quadric.

Proof. It remains to determine double points. We start from the following geometric description of $\Gamma(E)$. The hypersurface $\Gamma(E)$ consists of points in $\mathbb{R}_1^4$ that define spheres tangent to the ellipse. All spheres touching in one point define an isotropic line in $\mathbb{R}_1^4$. If two isotropic lines have a common point then this point is a double point on $\Gamma(E)$. Therefore, a common point of two isotropic lines corresponds to a sphere that touches the ellipse E in two points. There are two families of circles on the (x, y)-plane that touch the ellipse E in two points. One family consists of inside circles with centers on the x-axis, another one of outside circles with centers on the y-axis. A pencil of spheres that contains this circle goes through each of such circle, i.e. spheres of the pencil are tangent to the ellipse E in two points. We notice that these two families of spheres are defined by two equations $y = 0$ and $x = 0$ in $\mathbb{R}_1^4$. Therefore, the equations of the hiperboloids H_1 and H_2 are obtained from the equation $F = 0$ as the hyperplane sections.

Remark 12. For the ellipse E in Theorem 11 hypersurface $\Gamma(E)$ has also other singularities (not only double points). Consider a projection of $\Gamma(E)$ to the (x, y)-plane. Singularities of the ellipse offsets define the evolute curve (see Fig. 5.8), which is an envelope of normal lines to the ellipse E in the plane. Let a surface $K \subset \Gamma(E)$ be an envelope of all isotropic lines in $\Gamma(E)$. Since the isotropic lines are projected to normals, K is projected to the evolute, and all points of K are singular in $\Gamma(E)$. The well-known parametrization of the evolute enables us to parametrize K:

$$K(t, s) = \left((b^2 - a^2)\cos^3 t,\; \frac{(b^2 + a^2)\sin^3 t}{a},\; h(t)\frac{1 - s^2}{2s},\; h(t)\frac{1 + s^2}{2s}\right),$$

where $h(t)^2 = (b^2 \sin^2(t) + a^2 \cos^2(t))^3/a^2$. Note that a general point on K belongs to a segment of an isotropic line on $\Gamma(E)$ bounded by hyperplanes $x = 0$ and

$y = 0$. This fact may be important for applications in order to avoid singular points on canal surfaces.

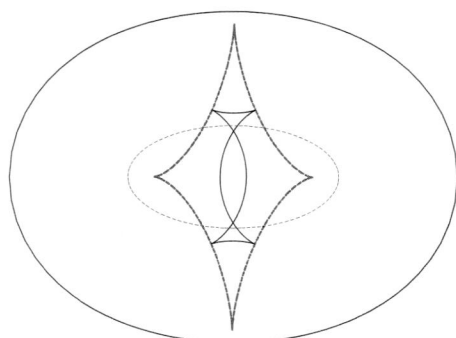

Fig. 5.8. The ellipse, two components of its offset and an evolute of the ellipse.

In other cases different degrees and different number of singular parts can appear. We will list only the most important information for blending applications: subsets of double points that contain positive lines.

For the case $\mathcal{P}_{++}^{(0)}$ represented by the parabola $C : \{x^2 - 4ay = 0, z = r = 0\}$ we calculate the following implicit equation of $\Gamma(C)$ of degree 6:

$$F = \omega^3 - 2\,v^2\omega^2 - \left(18\,ax^2 - v^3\right) v\omega - a\left(27\,ax^2 - 2\,v^3\right) x^2,$$

where $v = y - 2a$, $\omega = x^2 + y^2 + z^2 - r^2$. The double hyperbolic paraboloid $H :$ $\{x = 0, 4\,y - 1 + 4\,z^2 - 4\,r^2 = 0\}$ is obtained by intersecting with the hyperplane $x = 0$.

The case $\mathcal{H}_{++}$ is defined by the curve

$$C(t) = (a(1 + t^2)/(2t), b(1 - t^2)/(2t), 0, 0).$$

Then, similarly to the case $\mathcal{E}_{++}$, $\deg \Gamma(C) = 8$ and just one double hyperboloid of 1-sheet $H : \{y = 0, x^2/(a^2 + b^2) + z^2/b^2 - r^2/b^2 = 1\}$ is found.

The case $\mathcal{H}_{+-}^{(++)}$ is defined by the curve

$$C(t) = (0, 0, 2at/(1 - t^2), b(1 + t^2)/(1 - t^2)).$$

Then $\deg \Gamma(C) = 8$ and two double hyperboloids $H_1 : \{z = 0, (x^2 + y^2)/a^2 - r^2/(a^2 - b^2) = 1\}$ and $H_2 : \{r = 0, (x^2 + y^2)/b^2 - z^2/(a^2 - b^2) = 1\}$ contain four families of positive lines. Note that H_2 is a usual hyperboloid of revolution in $\mathbb{R}^3$.

The case $\mathcal{H}_{+0}$ is defined by the hyperbola

$$C(t) = (2at/(1 - t^2), 0, b\left(1 + t^2\right)/(1 - t^2), b\left(1 + t^2\right)/(1 - t^2))\}.$$

We have $\deg \Gamma(C) = 8$ and one double hyperboloid

$$\{x = 0, \left(b^2 - a^2\right) z^2 - 2 b^2 rz + \left(a^2 + b^2\right) r^2 + a^4 - a^2 y^2 = 0\}.$$

The case $\mathcal{H}^{[1]}_{+-}$ corresponds to the hyperbola $C(t) = (0, 0, 2at/(1 - t^2) . a(1 + t)^2/(1 - t^2))$, and $\deg \Gamma(C) = 6$. Here the two families of lines are lying on the hyperboloid $\{z = r, x^2 + y^2 - z^2 = a^2/2\}$ which is not double. This is caused by the following fact: natural projections from C to all of these lines are 1–1 but not 2–1 as in all previous cases, since one asymptote of C is isotropic.

5.7 Conclusions

Quadratic canal surfaces are natural generalizations of Dupin cyclides with the potential applications in geometric modeling, since

- they have a relatively simple rational parametrization of bi-degree $(3, 2)$ or $(4, 2)$;
- their Bézier representation is invariant with respect to Laguerre transformations (in particular, offsets have Bézier representations of the same bi-degree);
- their implicit degree is 6 or 8;
- they are tangent with families of circular cones and cylinders (also along non-circular curves).

In Fig. 5.9 we see three blendings between natural quadrics . The first two use $(4, 2)$-patches of a quadratic canal surface of type $\mathcal{E}_{++}$ for blending a cone and a cylinder. The third uses the biangle patch of bi-degree $(6, 2)$ of the same surface as fixed radius rolling ball blend of two cylinders with a common inscribed sphere.

Fig. 5.9. Blendings with patches of canal surfaces of type $\mathcal{E}_{++}$.

References

1. Degen, W., Cyclides, in Handbook of Computer Aided Geometric Design, 2002, p.575–601

2. M. Kazakevičiūtė, Classification of pairs of natural quadrics from the point of view of Laguerre geometry, *Lithuanian Mathematical Journal*, **45** (2005), 63–84.
3. Krasauskas, R., Minimal rational parametrizations of canal surfaces, *Computing*, to appear.
4. Krasauskas, R. and M. Kazakevičiūtė, Universal rational parametrizations and spline curves on toric surfaces, in: *Computational Methods for Algebraic Spline Surfaces*, ESF Exploratory Workshop, Springer, 2005, pp. 213–231.
5. Krasauskas, R. and Mäurer, C., Studying cyclides using Laguerre geometry, *Computer Aided Geometric Design* **17** (2000) 101–126.
6. Landsmann, G., Schicho J., and Winkler, F., The parametrization of canal surfaces and the decomposition of polynomials into a sum of two squares, *J. Symbolic Computation* **32** (2001) 119–132.
7. Peternell, M. and Pottmann, H., Computing rational parametrizations of canal surfaces, *J. Symbolic Computation* **23** (1997), 255–266.
8. Pottmann, H. and Peternell, M., Application of Laguerre geometry in CAGD, *Computer Aided Geometric Design* **15** (1998), 165–186.
9. Pratt, M. J., Cyclides in computer aided geometric design, *Computer Aided Geometric Design* **7** (1990), 221–242.
10. Zube, S., Bidegree $(2, 1)$ parametrizable surfaces in projective 3-space, *Lithuanian Math. J.* **38** (1998), 379–402.

6

General Classification of (1,2) Parametric Surfaces in $\mathbb{P}^3$

Thi-Ha Lê and André Galligo

Laboratoire J-A. Dieudonné
Université de Nice Sophia-Antipolis
Parc Valrose, 06108 Nice Cedex 2, France
{lethiha, galligo}@math.unice.fr

Summary. Patches of parametric real surfaces of low degrees are commonly used in Computer Aided Geometric Design and Geometric Modeling. However the precise description of the geometry of the whole real surface is generally difficult to master, and few complete classifications exist.

Here we study surfaces of bidegree (1,2). We present a classification and a geometric study of parametric surfaces of bidegree (1,2) over the complex field and over the real field by considering a dual scroll. We detect and describe (if it is not void) the trace of self-intersection and singular locus in the system of coordinates attached to the control polygon of a patch (1,2) in the box $[0; 1] \times [0; 1]$.

6.1 Introduction

We consider a polynomial mapping of bidegree (1,2):

$$\Phi : \mathbb{P}^1 \times \mathbb{P}^1 \longrightarrow \mathbb{P}^3$$

given by a matrix $\mathcal{A} = (a_{ij})$, $i = 1, \ldots, 4$, $j = 1, \ldots, 6$ of maximal rank 4 such that:

$$\Phi = {}^t(\Phi_1, \Phi_2, \Phi_3, \Phi_4) \text{ and } \Phi = \mathcal{A}.{}^t(tu^2, tuv, tv^2, su^2, suv, sv^2) \quad (1)$$

where$((t : s), (u : v))$ are a system of coordinates of $\mathbb{P}^1 \times \mathbb{P}^1$. The base field is $K = \mathbb{C}$ or $\mathbb{R}$. Then $\mathcal{S} = Im(\Phi) \subset \mathbb{P}^3$ is a parametric surface of bidegree (1,2) and Φ is a parameterization of $\mathcal{S}$.

Similarly, one defines surfaces of bidegree (m, n); patches of these surfaces are often used in C.A.G.D and Solid Modeling especially for the bi-cubics $m = n = 3$.

Our aim is to classify the applications Φ of bidegree (1,2) while the base field is $\mathbb{R}$ or $\mathbb{C}$ up to a change of projective coordinates in the source space $\mathbb{P}^1 \times \mathbb{P}^1$ and in the target space $\mathbb{P}^3$. In a previous article [13] we described the generic complex case and the geometry of the corresponding surfaces. Then, the parameterization of Φ is equivalent to a parameterization, we called "normal form" and denoted by $NF(a, b)$:

$$X = tu^2, Y = (t - s)(u - v)^2, Z = (t - as)(u - bv)^2, T = sv^2$$

where a and b are two complex parameters different from 0 and 1. Moreover, if $(a, b) \neq (a', b')$ then $NF(a, b)$ is not equivalent to $NF(a', b')$. We say that (a, b) is a couple of moduli for this classification.

In this article we study the real generic cases and the non generic cases. The surfaces $\mathcal{S}$ defined by (1) are ruled surfaces which admit an implicit equation in $\mathbb{P}^3$ of degree at most 4. These surfaces were studied extensively in the 19th century by great mathematicians: Cremona [7], Cayley [3], [4], Segre [28]; one finds a synthesis of theirs results and extensions in the books of Salmon [25] and of Edge [12]. From 1930, the main stream of algebraic geometry concentrated on the study of varieties up to birational equivalence and with more conceptual (and less effective) tools. However, applications in C.A.G.D and Solid Modeling showed the necessity of revisiting of the geometry of parametrized curves and surfaces of small degrees and bidegrees. An article of Coffman and al. [6] is a model of this kind of work: it revisited and completed the classification of parameterized surfaces of total degree 2 (started by Steiner in 1850). The ruled surfaces of implicit degree 4 are more complicated and have more diversity. In the 19th century the focus was not on the classification of parameterizations but rather on the geometric property and the calculation of certain invariants as well as on the obtaining of lists of implicit equations which are dependent of many parameters. A presentation of these classification results over the complex field related to rational (1,2)-Bézier surfaces with the description of the behaviour in presence of base-points, but without any description of the singularities, was provided by W.L.F. Degen [9]. A more complete classification over the real field, describing also the possible singularities was provided by S. Zube in [30] and [31]. Here we briefly review all these results, then we provide a new presentation based on the study of the dual scroll and the consideration of the tangent planes to all conics of the surface. We provide normal forms of the parameterizations and relate them to geometric data of the surface. We also consider the problem of defining classifying spaces which express the proximity with respect to deformations of these objects. Our article is organized as following:

In section 2, we recall some results of the 19th century, we follow the syntax given by Edge [12] in 1931, then we distinguish different types of parametric surfaces and we concentrate on the surfaces of bidegree (1,2). In section 3, we present our method of classification and introduce a scroll surface in the dual space which will be used to find the moduli. This variety is different from those used by the geometricians of 19th century but similar to the ones considered in [23]. In section 4, we recall the results obtained in [13] for generic complex case and extend them to the real setting. In section 5, we classify the intersections of a scroll (1,2) of $\mathbb{P}^5$ and a 3-projective plane or equivalently to the intersections of two curves of bidegree (1,2) in $\mathbb{P}^1 \times \mathbb{P}^1$. Then we apply these results to the classification of parametric surfaces (1,2). In section 6, we provide simple formulae to describe the critical points in the system of coordinates attached to the control polygon of a patch (1,2). We detect and describe the trace of the pre-images of the self-intersection and singular locus in the box $[0, 1] \times [0, 1]$.

6.2 Rational ruled surfaces

6.2.1 Ruled surfaces

Definition 1. *A **ruled surface** in projective space is a surface formed by a 'singly infinite system" of straight lines. The lines are called the generators of the surface.*

*A **normal ruled surface** is a ruled surface which can not be obtained by projection from another ruled surface of the same degree in a space of higher dimension.*

Proposition 2. *A rational ruled surface of degree n spanning $\mathbb{P}^{n+1}$ is normal in $\mathbb{P}^{n+1}$.*

All rational ruled surfaces of degree n can be obtained as projections of these normal surfaces.

Proof. See [12], pp 34-36.

Hence, a quartic rational ruled surface S in $\mathbb{P}^3$ can be obtained as projection of a rational normal quartic ruled surface $\mathcal{F}$ in $\mathbb{P}^5$. The center of projection is a line L. See also [9] for a classification of the relative positions of $\mathcal{F}$ and L, while in [30] the projection is decomposed into a projection on $\mathbb{P}^4$ followed by a projection on $\mathbb{P}^3$ in order to better describe the provided classification.

6.2.2 Directrices of a surface

We assume that $\mathcal{F}$ is not a cone (this case is simple).

Definition 3. *A **directrix curve** of a ruled surface is a curve on the surface meeting every generator in one point.*

*A **minimum directrix** is a directrix curve which is of minimum degree on the surface.*

Remark 4. The image of a directrix (respectively generator) of $\mathcal{F}$ by projection is a directrix (respectively generator) of S. Moreover, the degree of a directrix of $\mathcal{F}$ is the same as the one of its image.

Proposition 5. *Let $\lfloor \frac{n}{2} \rfloor$ denote $\frac{n}{2}$ if n is even and $\frac{n-1}{2}$ if n is odd. There are $\lfloor \frac{n}{2} \rfloor$ projectively distinct types of rational normal ruled surfaces of degree n in $\mathbb{P}^{n+1}$, each one has a directrix of minimum degree m, where $m = 1, 2, \ldots, \lfloor \frac{n}{2} \rfloor$.*

Proof. See [12], pp 38-39.

For $n = 4$ there are only two types, either with minimum directrix conics or with minimum directrix lines.

6.2.3 Parametric surfaces of bidegree $(1, 2)$

Definition 6. *Parametric surfaces of bi-degree (1,2) are images of maps*

$$\Phi : \mathbb{P}^1(\mathbb{C}) \times \mathbb{P}^1(\mathbb{C}) \longrightarrow \quad \mathbb{P}^3(\mathbb{C})$$
$$([t : s], [u : v]) \longmapsto [\Phi_1 : \Phi_2 : \Phi_3 : \Phi_4]$$

where $\Phi_1, \Phi_2, \Phi_3, \Phi_4$ are bihomogeneous polynomials in $[t : s]$ and $[u : v]$ of bidegree $(1, 2)$.

The parametric surfaces of bidegree $(1, 2)$ are rational ruled surfaces and have implicit degree 4 if $\Phi_1, \Phi_2, \Phi_3, \Phi_4$ have no base points. These $(1, 2)$ parametric surfaces $\mathcal{S}$ are images of the normal surfaces $\mathcal{F}$ having minimum directrix conics and generated by $(1, 1)$ correspondence between 2 directrix conics. Hence, $\mathcal{S}$ is generated by $(1, 1)$ correspondence between two non degenerated conics, or between a double line and a conic, or between two double lines (respectively when the center of projection is in general position in regard to $\mathcal{F}$, or it cuts a plane containing a directrix conic of $\mathcal{F}$ or it cuts two planes containing two directrix conics of $\mathcal{F}$.

The implicit equations for each case are given in [12], pp (62-69). These equations contain many parameters. We aim to consider normal forms for bidegree (1,2) parameterizations with a minimum number of parameters in the complex and in the real settings.

We denote by $(X : Y : Z : T)$ projective coordinates in $\mathbb{P}^3$ and by $(X : Y : Z : T : P : Q)$ projective coordinates in $\mathbb{P}^5$.

6.3 Dual scroll

We write the $(1, 2)$ parametric surfaces $\mathcal{S}$ in the basis

$$\{tu^2, 2tuv, tv^2, su^2, 2suv, sv^2\},$$

$$(\mathcal{S}) : \begin{cases} X = a_1 tu^2 + 2b_1 tuv + c_1 tv^2 + d_1 su^2 + 2e_1 suv + f_1 sv^2 \\ Y = a_2 tu^2 + 2b_2 tuv + c_2 tv^2 + d_2 su^2 + 2e_2 suv + f_2 sv^2 \\ Z = a_3 tu^2 + 2b_3 tuv + c_3 tv^2 + d_3 su^2 + 2e_3 suv + f_3 sv^2 \\ T = a_4 tu^2 + 2b_4 tuv + c_4 tv^2 + d_4 su^2 + 2e_4 suv + f_4 sv^2 \end{cases} \qquad (6.1)$$

where $a_i, b_i, c_i, d_i, e_i, f_i \in \mathbb{C}$.

Notation: $\mathcal{A}$ is the 4×6 matrix of the coefficients $a_i, b_i, c_i, d_i, e_i, f_i$. We can assume that $rank(\mathcal{A}) = 4$.

The considered surface $\mathcal{S}$ can be seen either as the total space of a family of conics $\mathcal{S} = \cup_t \mathcal{C}_t$ with $t \in \mathbb{P}^1(\mathbb{C})$, or as the total space of a family of lines $\mathcal{S} = \cup_u \mathcal{L}_u$ with $u \in \mathbb{P}^1(\mathbb{C})$.

6.3.1 A 3-projective plane

Definition 7. *We consider the map between dual spaces:*

$$\pi_{\mathcal{A}} : (\mathbb{P}^3)^* \rightarrow (\mathbb{P}^5)^*$$
$$(\alpha, \beta, \gamma, \delta) \mapsto (A, C, E, B, D, F) = (\alpha, \beta, \gamma, \delta)\mathcal{A}$$

defined by $^t\mathcal{A}$. *Its image is a 3-projective plane in* $(\mathbb{P}^5)^*$ *that we denote by* $\Pi_{\mathcal{A}}$.

By linear transformation, we can write the implicit equations of $\Pi_{\mathcal{A}}$ in $(\mathbb{P}^5)^*$ as follows:

$$\begin{cases} A_1X_1 + B_1X_2 + C_1X_3 + D_1X_4 + E_1X_5 + F_1X_6 = 0 \\ \quad\quad B_2X_2 + C_2X_3 + D_2X_4 + E_2X_5 + F_2X_6 = 0 \end{cases}$$

where $(X_1 : X_2 : X_3 : X_4 : X_5 : X_6)$ are projective coordinates of $(\mathbb{P}^5)^*$.

6.3.2 Tangent planes to all conics of the surface

We want to characterize the planes Π in $\mathbb{P}^3$ such that Π is tangent to any curve $\mathcal{C}_{(t:s)}$ of $\mathcal{S}$ or contains it.

The general equation of a plane Π in $\mathbb{P}^3$ is:

$$\alpha X + \beta Y + \gamma Z + \delta T = 0 \quad\quad (\alpha, \beta, \gamma, \delta) \in \mathbb{C}^4 \setminus \{0\} \quad\quad (6.2)$$

Substituting in (6.2) the expressions of X, Y, Z, T given in (6.1), we obtain the equation of the intersection of Π and of a conic $\mathcal{C}_{(t:s)}$:

$$\Pi \cap \mathcal{C}_{(t:s)} : (At + Bs)u^2 + 2(Ct + Ds)uv + (Et + Fs)v^2 = 0.$$

where $(A, C, E, B, D, F) = \pi_{\mathcal{A}}(\alpha, \beta, \gamma, \delta) \in (\mathbb{P}^5)^*$.

They are tangent (or Π contains $\mathcal{C}_{(t:s)}$) for all $(t : s) \in \mathbb{P}^1$ if and only if the discriminant vanishes identically, i.e. $(Ct + Ds)^2 - (At + Bs)(Et + Fs) = 0, \forall (t : s) \in \mathbb{P}^1$. This is true if and only if the following conditions are satisfied:

$$C^2 = AE \ , \quad 2CD = AF + BE \ , \quad D^2 = BF.$$

From this, we obtain four simpler equations:

$$C^2 = AE \ , \ D^2 = BF \ , \ CD = AF = BE \ , \quad\quad (6.3)$$

(We note that the four equations above are related). We have:

$$(6.3) \Leftrightarrow \text{rank} \begin{pmatrix} A & C & B & D \\ C & E & D & F \end{pmatrix} \leq 1. \quad\quad (6.4)$$

(6.4) defines a surface of $(\mathbb{P}^5)^*$, we denote by $\mathbb{F}(2,2)^*$ (which is a so-called *rational scroll*. So we have the following proposition:

Proposition 8. *A plane Π defined by $(\alpha, \beta, \gamma, \delta)$ in $\mathbb{P}^3$ is tangent to all conics of $\mathcal{S}$ (or contains) it if and only if $\pi_A(\alpha, \beta, \gamma, \delta) \in \mathbb{F}(2,2)^*$.*

In the following section, we express $\mathbb{F}(2,2)^*$ as the dual scroll (in a geometric sense that we will make precise) of the scroll $\mathbb{F}(2,2)$ and construct related parametric equations for $\mathbb{F}(2,2)^*$.

6.3.3 Parameterization of the dual scroll

Notations: We use affine coordinates t instead of $(t : s)$, u instead of $(u : v)$. We set the following notation and parametric equations of the scroll $\mathbb{F}(2,2)$ in $\mathbb{P}^5$ (it is the normal ruled surface of bidegree (1,2)):

$$\mathbb{F}(2,2) : \begin{cases} X = tu^2 \\ Y = tu \\ Z = t \\ T = u^2 \\ P = u \\ Q = 1 \end{cases}$$

$\mathbb{F}(2,2)$ is a surface and not a hypersurface. However for each point of $\mathbb{F}(2,2)$ we want to construct a hyperplane naturally attached to that point (This process is somehow similar to the construction of the osculating plane attached to a point of a space curve). These hyperplanes will describe a projective variety that we call the "dual scroll" in $(\mathbb{P}^5)^*$. This is not related to the usual but to a generalized notion of duality, already studied in [23] and called "strict duality".

Construction: We consider the affine chart $Q = 1$ where $\mathbb{F}(2,2)$ becomes an affine complete intersection, then its affine implicit equations are:

$$\begin{cases} X - TZ = 0 \\ Y - ZP = 0 \\ T - P^2 = 0 \end{cases}$$

We denote by M the parameterization map of the scroll $\mathbb{F}(2,2)$. To each point $M_0 = M(t_0, u_0)$, $((t_0, u_0) \neq (0,0))$ of the scroll, we associate generalized tangent spaces of dimension 3 and 4 constructed from $\mathcal{L}_{u_0}$ and $\mathcal{C}_{t_0}$ (that are the generator and the conic of the scroll passing through M_0).

The parametric equations of the line $\mathcal{L}_{u_0}$ and of the conic $\mathcal{C}_{t_0}$ are:

$$\mathcal{L}_{u_0} : \begin{cases} X = tu_0^2 \\ Y = tu_0 \\ Z = t \\ T = u_0^2 \\ P = u_0 \end{cases} \qquad \mathcal{C}_{t_0} : \begin{cases} X = t_0u^2 \\ Y = t_0u \\ Z = t_0 \\ T = u^2 \\ P = u \end{cases}$$

Therefore, we deduce implicit equations of plane Π_{t_0} containing C_{t_0}:

$$\Pi_{t_0} : \begin{cases} X - t_0 T = 0 \\ Y - t_0 P = 0 \\ Z - t_0 \ = 0. \end{cases}$$

The intersection of $\mathcal{L}_{u_0}$ and Π_{t_0} is just the point M_0. We denote by $\mathcal{G}(t_0, u_0)$ the affine space generated by $\mathcal{L}_{u_0}$ and Π_{t_0}. Hence, $\mathcal{G}(t_0, u_0)$ has dimension 3. Implicit projective equations of $\mathcal{G}(t_0, u_0)$ are:

$$\mathcal{G}(t_0, u_0) : \begin{cases} X - t_0 T - u_0^2 Z + u_0^2 t_0 Q = 0 \\ Y - u_0 Z - t_0 P + u_0 t_0 Q = 0. \end{cases}$$

We denote by $\mathcal{E}$ the set of 4-projective spaces containing $\mathcal{G}(t_0, u_0)$; such a hyperplane is denoted by $H(\alpha, \beta, t_0, u_0)$ and have an equation of type:

$$\alpha(X - t_0 T - u_0^2 Z + u_0^2 t_0 Q) + \beta(Y - u_0 Z - t_0 P + u_0 t_0 Q) = 0$$

where $(\alpha, \beta) \in \mathbb{C}^2 - \{0\}$.

Each hyperplane $H(\alpha, \beta, t_0, u_0)$ cuts the scroll along a curve of degree 4 (because the scroll has degree 4). As it already contains C_{t_0} and $\mathcal{L}_{u_0}$, the intersection must contain another line of the scroll, let us call it $\mathcal{L}_{u'}$. We aim to single out the hyperplane $H(\alpha, \beta, t_0, u_0)$ such that $\mathcal{L}_{u_0} \equiv \mathcal{L}_{u'}$. By replacing the parametric expressions of X, Y, Z, T, P, Q of $\mathbb{F}(2, 2)$ in the equation above, we obtain the equation of the intersection of $\mathbb{F}(2, 2)$ and $H(\alpha, \beta, t_0, u_0)$ in the parameter space of the scroll:

$$(u - u_0)(t - t_0)(\alpha u + \alpha u_0 + \beta) = 0$$

Therefore, $u' = \dfrac{-(\alpha u_0 + \beta)}{\alpha}$. ($\alpha$ must be different from 0, otherwise $\alpha = \beta = 0$).

We get, $u' = u_0$ if and only if: $\dfrac{-(\alpha u_0 + \beta)}{\alpha} = u_0 \Rightarrow \beta = -2\alpha u_0$. We can take $\alpha = 1$, so we have: $\beta = -2u_0$. In this case $H(1, -2u_0, t_0, u_0)$ (denoted by $H(t_0, u_0)$) cuts $\mathbb{F}(2, 2)$ in C_{t_0} and twice in $\mathcal{L}_{u_0}$. The equation of $H(t_0, u_0)$ becomes:

$$X - 2u_0 Y + u_0^2 Z - t_0 T + 2u_0 t_0 P - u_0^2 t_0 Q = 0.$$

The coefficients are the coordinates $(A_1 : \ldots : A_6)$ of this hyperplane in $(\mathbb{P}^5)^*$ in the following order:

$$\begin{pmatrix} 1 & -2u_0 & u_0^2 & -t_0 & 2u_0 t_0 & -u_0^2 t_0 \\ A_1 & A_2 & A_3 & A_4 & A_5 & A_6 \end{pmatrix}.$$

and satisfy the condition:

$$\text{rank} \begin{pmatrix} A_1 & \dfrac{1}{2}A_2 & A_4 & \dfrac{1}{2}A_5 \\ \dfrac{1}{2}A_2 & A_3 & \dfrac{1}{2}A_5 & A_6 \end{pmatrix} = 1.$$

This condition defines a scroll $\mathbb{F}(2,2)^*$ in $(\mathbb{P}^5)^*$ that we call the dual scroll of $\mathbb{F}(2,2)$. The parametric equations of the scroll $\mathbb{F}(2,2)^*$ in the dual space $(\mathbb{P}^5)^*$ were given above.

6.3.4 Intersection of $\varPi_A$ and $\mathbb{F}(2,2)^*$

By replacing the parametric equations of $\mathbb{F}(2,2)^*$ in the implicit equation of $\varPi_A$, we see that $\varPi_A \cap \mathbb{F}(2,2)^*$ is given by the intersection of two curves of bidegree (1,2) in the parameter space $\mathbb{P}^1 \times \mathbb{P}^1$:

$$\begin{cases} \varphi_1(t,u) = A_1 - 2B_1u + C_1u^2 - D_1t + 2E_1tu - F_1u^2t = 0 \\ \varphi_2(t,u) = 2B_2u + C_2u^2 - D_2t + 2E_2tu - F_2u^2t = 0. \end{cases}$$

We have two cases: either $\varphi_1(t,u) \cap \varphi_2(t,u)$ is finite (4 points) or infinite. We first consider the generic cases, i.e. the intersection contains 4 distinct points (t_k, u_k), $k = 1, \ldots, 4$ and $t_k \neq t_j$, $u_k \neq u_j$ if $k \neq j$. This will give a classification of the maps of bidegree (1,2) up to change of coordinates and a set of normal forms.

6.4 The generic case

We first recall the result for the generic complex case (see more details in the article [13]).

6.4.1 The generic complex case

Generically, $\varphi_1(t,u) \cap \varphi_2(t,u)$ contains 4 distinct points; they correspond to 4 tangent planes. They are tangent to all conics of $\mathcal{S}$, along a special torsal line.

We can choose these 4 tangent planes in $\mathbb{P}^3$ to be the planes of coordinates $(X = 0), (Y = 0), (Z = 0), (T = 0)$.

We proved in [13] that, after a suitable change of coordinates and change of parameters, the surface $\mathcal{S}$ admits the parametric representation:

$$\begin{cases} X = tu^2 \\ Y = (t-s)(u-v)^2 \\ Z = (t-as)(u-bv)^2 \\ T = sv^2 \end{cases}$$

This normal form depends on two moduli a and b. Moreover, the singular locus of the surface is a twisted cubic and has parametric equations (c.f [13]): $(t : s) \longmapsto (X : Y : Z : T)$

$$\begin{cases} X = abt(bt - t - bs + as)^2 \\ Y = (a-1)(as - bt)(b^2t^2 - b^2t^2 + b^2t + bats - bts - ats + as^2 - bas^2) \\ Z = a(a-1)b(bs - t)(b^2t^2 - t^2 + bts + ats - bats - b^2ts - bas^2 + b^2as^2) \\ T = (at - bt - as + bas)^2s \end{cases}$$

Each singular point $(X(t : s), Y(t : s), Z(t : s), T(t : s))$ is the intersection of two lines $\mathcal{L}_{(u_1:v_1)}$ and $\mathcal{L}_{(u_2:v_2)}$ which belong to the plane $\Pi_{(t:s)}$ containing the conic $\mathcal{C}_{(t:s)}$ and to the surface, i.e. $\Pi_{(t:s)} \cap S = \mathcal{C}_{(t:s)} \cup \mathcal{L}_{(u_1:v_1)} \cup \mathcal{L}_{(u_2:v_2)}$, where $(u_1 : v_1)$, $(u_2 : v_2)$ are roots of the equation:

$$[(b-a)t+a(1-b)s]su^2+2b(a-1)tsuv+(b-b^2)t^2v^2+(b-a)btsv^2 = 0. \quad (6.5)$$

In the sequel, in order to simplify the readability, we shall often use affine coordinates t instead of $(t : s)$, u instead of $(u : v)$ and so on.

6.4.2 The generic real case

Generically, the intersection of $\varphi_1(t, u)$ and $\varphi_2(t, u)$ is 4 distinct points:

$$(t_1; u_1), \ (t_2; u_2), \ (t_3; u_3), \ (t_4; u_4),$$

moreover all the t_i (and all the u_i) are two by two distinct. These four points correspond to four special tangent planes.

We observe that as the equations $\varphi_1(t, u)$ and $\varphi_2(t, u)$ have degree 1 in t, if u is real then t is also real; and if u_1, u_2 are complex conjugate then the same holds for t_1 and t_2. So we have 3 cases that we denote by type I, type II, type III: either 4 real points, or 2 real points and 2 conjugate points or two couples of conjugate points. For all types, as in the generic complex case, each singular point is intersection of two lines $\mathcal{L}_{u_1}$ and $\mathcal{L}_{u_2}$ where u_1, u_2 are the roots of an equation of degree 2 whose coefficients are real polynomials in t. So either u_1, u_2 are reals or conjugate complex. Therefore $\mathcal{L}_{u_1}$ and $\mathcal{L}_{u_2}$ are two conjugate lines. Their intersection is always real. Hence, the singularity of the complex surface is real and moreover is a twisted cubic. However only segments of this real twisted cubic form the singular locus of the real parametric surface.

The study of the first case (type I) is as in the generic complex case. We present the two last cases.

a) Two real and two conjugate points: type II

Lemma 9. *We assume that* $t_1, t_2, u_1, u_2 \in \mathbb{R}$ *et* $t_3, t_4, u_3, u_4 \in \mathbb{C}$ *and* $t_3 = \bar{t}_4$ *and* $u_3 = \bar{u}_4$. *Hence, it exists two real homographies:* $\eta_1, \eta_2 : \mathbb{P}^1(\mathbb{R}) \rightarrow \mathbb{P}^1(\mathbb{R})$ *and two values* $\theta, \theta' \in [0, \pi]$ *such that:*

$$\eta_1(t_1) = 0, \eta_1(t_2) = \infty, \eta_1(t_3) = e^{i\theta}, \eta_1(t_4) = e^{-i\theta}$$

$$\eta_2(u_1) = 0, \eta_2(u_2) = \infty, \eta_2(u_3) = e^{i\theta'}, \eta_2(u_4) = e^{-i\theta'}$$

The proof is simple but tedious.

Therefore, by choosing 4 tangent planes as $(X = 0), (Y = 0), (Z = 0), (T = 0)$ and by a similar demonstration as in the generic complex case, we obtain the parametric complex representation of the surface:

$$\begin{cases} X = tu^2 \\ Y = (t - e^{i\theta}s)(u - e^{i\theta'}v)^2 \\ Z = (t - e^{-i\theta}s)(u - e^{-i\theta'}v)^2 \\ T = sv^2 \end{cases}$$

We write the surface equations in the affine chart $s = v = T = 1$ and take we have that:

$$\begin{cases} x = tu^2 \\ y = (t - \cos\theta - i\sin\theta)(u - \cos\theta' - i\sin\theta')^2 \\ z = (t - \cos\theta + i\sin\theta)(u - \cos\theta' + i\sin\theta')^2 \end{cases}$$

By dividing y and z by $\sin\theta\sin^2\theta'$ and denoting $a = \cotan\theta$, $b = \cotan\theta$, we obtain the following system:

$$\begin{cases} x = tu^2 \\ \dfrac{y}{\sin\theta\sin^2\theta'} = (\dfrac{t}{\sin\theta} - a - i)(\dfrac{u}{\sin\theta'} - b - i)^2 \\ \dfrac{z}{\sin\theta\sin^2\theta'} = (\dfrac{t}{\sin\theta} - a + i)(\dfrac{u}{\sin\theta'} - b + i)^2 \end{cases}$$

By changing the parameters $t' = \dfrac{t}{\sin\theta} - a$, $u' = \dfrac{u}{\sin\theta'} - b$ and by transformation of coordinates $(x', y', z') = \dfrac{1}{\sin\theta\sin^2\theta'}(x, y, z)$ we obtain the surface equations as follows:

$$\begin{cases} x' = (t' + a)(u' + b)^2 \\ y' = (t' - i)(u' - i)^2 \\ z' = (t' + i)(u' + i)^2 \end{cases}$$

Finally, we transform $(x'', y'', z'') = (x', \dfrac{y' + z'}{2}, \dfrac{y' - z'}{-2i})$. Therefore we proved the following proposition:

Proposition 10. *A normal form for the parametric equations of a surface of type II is as follows:*

$$(\mathcal{S}) : \begin{cases} x = (t + a)(u + b)^2 \\ y = tu^2 - t - 2u \\ z = 2tu + u^2 - 1 \end{cases} \quad \text{with } a = \cotan\theta, b = \cotan\theta'$$

A surface of type II has two real pinch points corresponding to $(t_1 = -b, u_1 = -b)$, $(t_2 = \infty, u_2 = \infty)$ and has two real torsal lines $\mathcal{L}_{u_1}$ and $\mathcal{L}_{u_2}$.

b) Two conjugate couples: type III

Lemma 11. *We assume that* $(t_1, u_1) = \overline{(t_2, u_2)}$ *and* $(t_3, u_3) = \overline{(t_4, u_4)}$. *It exists two real homographies* $\eta_1, \eta_2 : \mathbb{P}^1(\mathbb{R}) \rightarrow \mathbb{P}^1(\mathbb{R})$ *and two values* $\theta, \theta' \in [0, \pi]$ *such that:*

$$\eta_1(t_1) = i, \eta_1(t_2) = -i, \eta_1(t_3) = e^{i\theta}, \eta_1(t_4) = e^{-i\theta}$$

$$\eta_2(u_1) = i, \eta_2(u_2) = -i, \eta_2(u_3) = e^{i\theta'}, \eta_2(u_4) = e^{-i\theta'}$$

Therefore, by choosing the four special tangent planes as $(X = 0), (Y = 0), (Z = 0), (T = 0)$ and by a similar demonstration as in the generic complex case, we have the parametric complex representation of the surface in the affine chart $s = v = 1$:

$$\begin{cases} X = (t - i)(u - i)^2 \\ Y = (t + i)(u + i)^2 \\ Z = (t - e^{i\theta})(u - e^{i\theta'})^2 \\ T = (t - e^{-i\theta})(u - e^{-i\theta'})^2 \end{cases}$$

By similar transformation as in the case (a), we obtain the following proposition (with two moduli θ and θ'):

Proposition 12. *A surface of type III has a real parameterization as follows:*

$$(\mathcal{S}) : \begin{cases} X = tu^2 - t - 2u \\ Y = 2tu + u^2 - 1 \\ Z = (\dfrac{t}{\sin\theta} - \cotan\theta)((\dfrac{u}{\sin\theta'} - \cotan\theta')^2 - 1) - 2(\dfrac{u}{\sin\theta'} - \cotan\theta') \\ T = 2(\dfrac{t}{\sin\theta} - \cotan\theta)(\dfrac{u}{\sin\theta'} - \cotan\theta') + (\dfrac{u}{\sin\theta'} - \cotan\theta')^2 - 1 \end{cases}$$

6.5 Non generic cases

We now list the following particular cases arising in the intersection of two curves of bidegree (1,2) whose equations are $\varphi_1(t, u)$ and $\varphi_2(t, u)$.

6.5.1 Their intersection is finite

Set $\varphi_1(t, u) \cap \varphi_2(t, u) = \{(t_1, u_1), (t_2, u_2), (t_3, u_3), (t_4, u_4)\}$. We distinguish the following cases:

    a) 4 distinct points.

        We have two cases: either $(t_1 = t_2$ and $t_3 \neq t_4)$ or $(t_1 = t_2$ and $t_3 = t_4)$.

    b) 2 distinct points and 1 double point $(t_3, u_3) = (t_4, u_4)$.

        We have 3 cases: either $(t_1 - t_2)(t_2 - t_3)(t_1 - t_3) \neq 0$, or $t_1 = t_2$ or $t_1 = t_3(= t_4)$.

    c) 2 double points.

    d) 1 triple point and 1 simple point.

6.5.2 Their intersection is infinite

$\varphi_1(t, u) \cap \varphi_2(t, u)$ is infinite if and only if $\varphi_1(t, u), \varphi_2(t, u)$ have a common factor, denoted by $g(t, u)$ and it is not constant. So we can write:

$$\begin{cases} \varphi_1(t, u) = g(t, u)\psi_1(t, u) \\ \varphi_2(t, u) = g(t, u)\psi_2(t, u) \end{cases}$$

We distinguish the following cases:
a) $g(t, u)$ is of bidegree (1,0).
b) $g(t, u)$ is of bidegree (1,1):
   $g(t, u)$ can be reduced.
   $g(t, u)$ cannot be reduced.
c) $g(t, u)$ is of bidegree (0,2):
   $g(t, u)$ has a double root.
   $g(t, u)$ has two different roots.
d) $g(t, u)$ is of bidegree (0,1).
   The system $\{\psi_1(t, u), \psi_2(t, u)\}$ has two different roots.
   The system $\{\psi_1(t, u), \psi_2(t, u)\}$ has a double root.

6.5.3 Parametric equations of the surface

We consider some particular cases and give the parametric equations of the surface for each case. The remaining cases can be treated similarly.

Remark 13. We remind that the 3-projective plane Π_A is defined by the transpose matrix of the matrix $\mathcal{A}$ of the parameterization of the surface. The equations of Π_A can be written:

$$\Pi_A : \begin{cases} A_1X_1 + B_1X_2 + C_1X_3 + D_1X_4 + E_1X_5 + F_1X_6 = 0 \\ A_2X_1 + B_2X_2 + C_2X_3 + D_2X_4 + E_2X_5 + F_2X_6 = 0 \end{cases}$$

We set:

$$\varphi_1 = (A_1, B_1, \ldots, F_1) \in \mathbb{C}^6 \setminus \{0\}$$

$$\varphi_2 = (A_2, B_2, \ldots, F_2) \in \mathbb{C}^6 \setminus \{0\}.$$

Therefore,

$$\Pi_A = \{X = {}^t(X_1, \ldots, X_6) \in \mathbb{C}^6 \setminus \{0\} \mid (\varphi_1, X) = (\varphi_2, X) = 0\}.$$

We observe that the rows of $\mathcal{A}$ are images of the points $(1 : 0 : \ldots : 0), \ldots (0 : \ldots : 1)$ by ${}^t\mathcal{A}$ so, they belong to Π_A. Hence $\ker \mathcal{A} = < {}^t\varphi_1, {}^t\varphi_2 >$.

If rank $\mathcal{A} = 4$, we can transform $\mathcal{A}$ to the echelon form:

$$A = \begin{pmatrix} 1 & 0 & 0 & 0 & \alpha_1 & \beta_1 \\ 0 & 1 & 0 & 0 & \alpha_2 & \beta_2 \\ 0 & 0 & 1 & 0 & \alpha_3 & \beta_3 \\ 0 & 0 & 0 & 1 & \alpha_4 & \beta_4 \end{pmatrix}.$$

Therefore $\ker A = < (\alpha_1 : \alpha_2 : \alpha_3 : \alpha_4 : -1 : 0), (\beta_1 : \beta_2 : \beta_3 : \beta_4 : 0 : -1) >$.
Hence if we know the equations of Π_A, we can deduce the matrix A and reversely.

a) The case (5.1.a) and $t_1 = t_2$:

By change of parameters, we can choose these four points as $(0,0)$, $(1,1)$, $(0,b)$ and (∞,∞). Hence, the parametric equations of the surface can be written as follows:

$$\begin{cases} X = tu^2 \\ Y = (t-s)(u-v)^2 \\ Z = t(u-bv)^2 \\ T = sv^2 \end{cases}$$

We observe that it is a limit situation of the generic case, namely where $a = 0$.

b) The case (5.1.b) and $(t_1 - t_2)(t_2 - t_3)(t_1 - t_3) \neq 0$:

We can choose 4 points as $(0,0)$, $(1,1)$, (∞,∞) where $(1,1)$ is double point. Therefore, the parametric equations of the surface can be written as follows:

$$\begin{cases} X = tu^2 \\ Y = (t-s)(u-v)^2 \\ Z = atu^2 + btuv + csu^2 + dtv^2 + esuv + fsv^2 \\ T = sv^2 \end{cases}$$

By linear transformation, in the affine chart $s = v = 1$, they are written:

$$\begin{cases} x = & tu^2 \\ y = & -2tu + t - u^2 + 2u \\ z = & btu + cu^2 + dt \end{cases}$$

If $b \neq 0$, we can take $b = 1$. From the surface equations above we deduce the equation of 3-projective plane Π_A:

$$\begin{cases} dX_2 - X_3 + (d + \dfrac{1}{2})X_5 = 0 \\ cX_2 - X_4 + (c - \dfrac{1}{2})X_5 = 0 \end{cases} \tag{6.6}$$

By replacing the expressions of X_2, X_3, X_4, X_5 of $\mathbb{F}(2,2)^*$ in (6.6) we obtain the equations of intersection of Π_A and $\mathbb{F}(2,2)^*$:

$$\begin{cases} -su(u + 2dv) + tu(2d + 1) = 0 \\ -2csuv + tv(v + (2c - 1)u) = 0 \end{cases}$$

$$\Longleftrightarrow \begin{cases} -2csuv + tv(v + (2c-1)u) = 0 \\ uv[(2c-1)u^2 + (1 - 2(d+c))uv + 2dv^2] = 0 \end{cases}$$

We have that, $(t = s = 1); (u = v = 1)$ is a double root of the system above if and only if $c = \dfrac{1}{2} + d, d \neq 0$. Therefore, the parametric equations of the surface is as follows:

$$(\mathcal{S}): \begin{cases} x = tu^2 \\ y = (t-1)(u-1)^2 \\ z = (d + \dfrac{1}{2})u^2 + tu + dt \end{cases} \qquad d \neq 0$$

If $b = 0$, we obtain the parametric equations of the surface:

$$(\mathcal{S}): \begin{cases} x = tu^2 \\ y = (t-1)(u-1)^2 \\ z = u^2 + t \end{cases}$$

c) The case (5.2.d) and the system $\{\psi_1(t, u), \psi_2(t, u)\}$ have two different roots:

We can write:

$$\begin{cases} \varphi_1(t, u) = g(t, u)\psi_1(t, u) = g(t, u)(tA_1(u) + A_2(u)) \\ \varphi_2(t, u) = g(t, u)\psi_2(t, u) = g(t, u)(tB_1(u) + B_2(u)) \end{cases}$$

where $A_1(u)$, $A_2(u)$, $B_1(u)$, $B_2(u)$ are polynomials of degree 1 in u.

We denote by u_0 the root of $g(t, u)$. We call (t_1, u_1) and (t_2, u_2) two roots of $\psi_1(t, u)$ and $\psi_2(t, u)$. We have two cases: either $u_1, u_2 \neq u_0$ or one of them is equal to u_0.

Firstly, we consider the case where $u_1, u_2 \neq u_0$. By change of parameters, we assume that $u_0 = 0$, $(t_1, u_1) = (1, 1)$, $(t_2, u_2) = (0, \infty)$. Hence, $\varphi_1(t, u)$ and $\varphi_2(t, u)$ become:

$$\begin{cases} \varphi_1(t, u) = u(tu - 1) \\ \varphi_2(t, u) = u(t - 1) \end{cases}$$

We deduce the equations of Π_A:

$$\Pi_A: \begin{cases} X_2 - 2X_6 = 0 \\ X_2 + X_5 \end{cases}$$

By the remark (13) we obtain the parametric equations of the surfaces:

$$(\mathcal{S}): \begin{cases} X = tu^2 \\ Y = u^2 \\ Z = 2tu - 2u + 1 \\ T = t \end{cases}$$

Then, if $u_1 = u_0 = 0$, one can choose $(t_1, u_1) = (1, 0)$, $(t_2, u_2) = (0, \infty)$. Similarly, we obtain the equations of the surface:

$$(\mathcal{S}) : \begin{cases} X = tu^2 \\ Y = u^2 \\ Z = tu - u \\ T = t \end{cases}$$

6.6 Detection of the singularities of a patch

Self-intersection curves of a polynomial patch are often computed approximately (see e.g. [8], [29], [20],...). Here we provide a symbolic method adapted to our setting.

We write the parametric equations of the surface in the Bernstein's basis

$$\{tu^2, t(1-u)^2, (1-t)u^2, (1-t)(1-u)^2, 2tu(1-u), 2(1-t)u(1-u)\}$$

and consider it in $[0,1] \times [0,1]$. The surface depends on the 6 control points, by changing coordinates we can choose these points to be: $(0 : 0 : 0 : 1)$, $(1 : 0 : 0 : 1)$, $(0 : 1 : 0 : 1)$, $(0 : 0 : 1 : 1)$, $(a : b : c : 1)$ and $(d : e : f : 1)$. Therefore the surface has a Bézier representation:

$$\begin{cases} x = t(1-u)^2 + 2atu(1-u) + 2d(1-t)u(1-u) \\ y = (1-t)u^2 + 2btu(1-u) + 2e(1-t)u(1-u) \\ z = (1-t)(1-u)^2 + 2ctu(1-u) + 2f(1-t)u(1-u) \end{cases}$$

We used the software Maple for our computation. We denote by F the implicit equation of the surface, F_x, F_y, F_z the partial derivatives of F and we denote by M_x, M_y, M_z polynomials in (t, u) obtained by replacing the parametric expressions of x, y, z in F_x, F_y, F_z. As F is of degree 4, M_x, M_y, M_z are of projective bedegree (3,6) but of affine bidegree (3,5) in (t, u). The implicit equation $\mathcal{C}(t, u)$ of the double locus (also denoted by $\mathcal{C}(t, u)$) in the parameter space of the surface is the gcd of M_x, M_y and M_z. The curve $\mathcal{C}(t, u)$ is a curve of degree (2,2). We have:

$$\frac{M_x}{\mathcal{C}(t, u)} = ((e + f - c - b - 1)u^2 + (2b - 2e + 1)u - b + e)t$$
$$+ (4ec - 4bf - 2c + 2b)u^3 + (1 + 2c - f - 6ec + 6bf - 4b - e)u^2$$
$$+ (2ec - 2bf - 1 + 2b + 2e)u - e$$

$$\frac{M_y}{\mathcal{C}(t, u)} = (c - f + a - d)(u - 1)^2 t + (4af - 2f - 4cd - 2a + 1)u^3 + (6cd - 6af$$
$$+ d - 3 + 5f + 4a)u^2 + (2af + 3 - 4f - 2d - 2a - 2cd)u + d + f - 1$$

$$\frac{M_z}{\mathcal{C}(t,u)} = ((e+a-1-d-b)u^2+(2b+1-2e)u-b+e)t+(4bd-1+2e+2a$$
$$- 4ea)u^3+(d-5e+2-2a+6ea-6bd)u^2+(2bd+4e-2ea-1)u-e.$$

By computation we check that the vector $[\dfrac{M_x}{\mathcal{C}(t,u)}, \dfrac{M_y}{\mathcal{C}(t,u)}, \dfrac{M_z}{\mathcal{C}(t,u)}]$ is equal to the

cross-product $\dfrac{\partial\Phi(t,u)}{\partial t} \wedge \dfrac{\partial\Phi(t,u)}{\partial u}$.

We consider the points (t,u) for which $\dfrac{\partial\Phi(t,u)}{\partial t} \wedge \dfrac{\partial\Phi(t,u)}{\partial u}$ vanishes, i.e the

common roots of $\dfrac{M_x}{\mathcal{C}(t,u)}, \dfrac{M_y}{\mathcal{C}(t,u)}, \dfrac{M_z}{\mathcal{C}(t,u)}$. More precisely, (t,u) is a root of the

system:

$$0 = (2af-e+1+2ea-2bf-f+b-c-2bd-2a+2ec-2cd)u^3 \qquad (6.7)$$
$$+ (f-2-5ec+5bd+2c-5ea+5bf-af-3b+3a+3e+cd)u^2$$
$$+ (1-3e-4bf+4ec-a-c-4bd+4ea+3b)u+e+bd+bf-ec-ea-b$$
$$t = \frac{((2a-4af+4cd+2f-1)u^2-(2cd+3f-2af+2a+d-2)u+d+f-1}{(a+c-d-f)(u-1)}$$

Generically, the system (6.7) has 4 roots (a root is (∞,∞)). They are the critical points of the parameterization and belong to the closure of the double locus $\mathcal{C}(t,u)$ in the parameter space. We denote them by E_1, E_2, E_3, E_4.

For each $t_0 \in \mathbb{P}^1$, we calculate the plane Π_{t_0} containing the conic $\mathcal{C}_{t_0}$ of the surface. The intersection $\Pi_{t_0} \cap S$ is determined by an equation $(t-t_0)g(t_0,u) = 0$, where $g(t_0,u)$ is polynomial of bidegree (2,2) in (t_0,u). If we consider it as polynomial of degree 2 in u so $g(t_0,u)$ has two roots u_1, u_2. Hence $\Pi_{t_0} \cap S = \mathcal{C}_{t_0} \cup \mathcal{L}_{u_1} \cup \mathcal{L}_{u_2}$. The polynomial $g(t_0,u)$ has a double root in u if its discriminant with respect to u vanishes and generically it vanishes for 4 values of t_0. For each one of these values we have a corresponding value of u depending on t_0 and being a double root of $g(t_0,u)$, hence we obtain 4 corresponding values of u (c.f [13]). Therefore the surface have 4 torsal lines corresponding to these four values of u. By replacing each of these values in the equation $\mathcal{C}(t,u)$ we obtain 4 critical points of the parameterization. They are actually the 4 points E_1, E_2, E_3, E_4. We denote their images by Φ respectively by P_1, P_2, P_3, P_4.

Hence, in the general case, the singular locus of S consists in a twisted cubic $\mathscr{C}$ (the closure of the image of the double locus $\mathcal{C}(t,u)$ of the parameterization) and 4 embedded points P_1, P_2, P_3, P_4 in $\mathscr{C}$. Near P_1 (or P_2, P_3, P_4) the surface is described by a continuous family of two lines $\mathcal{L}_{u_1}$ and $\mathcal{L}_{u_2}$ intersecting in a point P of $\mathscr{C}$. But when $P \equiv P_1$ the lines coincide and we obtain a torsal line of the surface.

By the implicit function theorem, we can apply a local isomorphism of $\mathbb{R}^3$ at P_i $(i=1,\ldots,4)$ which transform locally the curve $\mathscr{C}$ into a line $(x_1 = 0, y_1 = 0)$

in the coordinates (x_1, y_1, z_1). The local equation of the surface becomes $x_1^2 - z_1 y_1^2$. Hence, P_i is a pinch point also called a **Whitney umbrella** singularity in the real setting. In our situation, the "rod of the umbrella" is curved, moreover the half line is not visible in the real parametric surface.

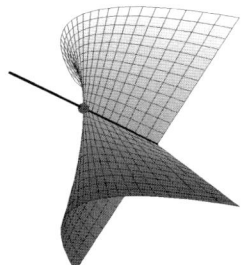

Fig. 6.1. Whitney umbrella

We have the following proposition:

Proposition 14. *1) A non degenerate cubic curve of $\mathbb{R}^3$ does not admit an oval.*

2) As a consequence, if the double locus in the parameter patch has an oval then this oval contain two critical points (i.e pre-images of two pinch points).

Proof. 1) It is easy to prove.

2) We recall that the parameterization map Φ restricted to $[0, 1]^2$ is continuous. If the double locus in the parameter space has an oval $\mathcal{O}$ (hence a compact set) its image by Φ is compact and is included in the singular locus $\mathscr{C}$. As $\mathscr{C}$ is a twisted cubic, it does not contain an oval, so the image of the oval must be a segment of curve delimited by two points P_1 and P_2. The pre-image of P_1 (respectively P_2) consists of only one point (a critical point of Φ) which belongs to $\mathcal{O}$. Hence, P_1 and P_2 are two of the pinch points of the surface.

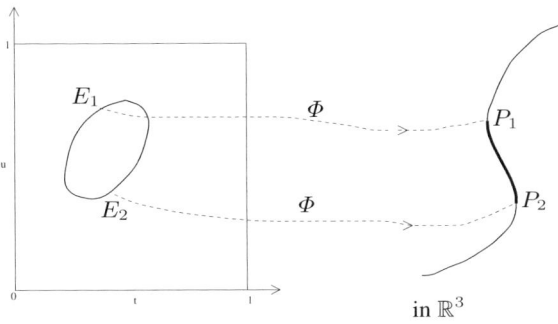

Corollary 15. *To detect an oval of the double locus in the box* $[0;1] \times [0;1]$*, it is sufficient to calculate the critical points of the parameterization. This provides an economy of calculation.*

Example 16. $a = 3$, $b = 5$, $c = 2$, $d = -4$, $e = 1$, $f = -2$.

The surface is defined by the parameterization:

$$\begin{cases} x = t(1-u)^2 + 6tu(1-u) - 8(1-t)u(1-u) \\ y = (1-t)u^2 + 10tu(1-u) + 2(1-t)u(1-u) \\ z = (1-t)(1-u)^2 + 4tu(1-u) - 4(1-t)u(1-u) \end{cases}$$

We have that:

$$\mathcal{C}(t,u) = 4(234476t^2u^2 - 446468t^2u - 215277tu^2 + 42550u^2 + 212531t^2 \\ + 496138tu - 131320u - 281358t + 92806)$$

$$\frac{M_x}{\mathcal{C}(t,u)} = (-9u^2 + 9u - 4)t - 86u^2 + 35u - 1 + 54u^3$$

$$\frac{M_y}{\mathcal{C}(t,u)} = (11 + 11u^2 - 22u)t - 7 + 7u^3 - 17u^2 + 17u$$

$$\frac{M_z}{\mathcal{C}(t,u)} = (2u^2 + 9u - 4)t - 85u^3 + 125u^2 - 43u - 1$$

Therefore the three affine critical points are the roots of the followed system:

$$\begin{cases} 73u^3 - 180u^2 + 148u - 39 = 0 \\ t = \frac{146}{11}u^2 - \frac{221}{11}u + \frac{85}{11} \end{cases}$$

The system above has only one root ($t_0 \approx 0.72427400$, $u_0 \approx 0.5442518227$) in the box $[0;1] \times [0;1]$, i.e the surface has only one critical point $E = (t_0, u_0)$ in the box $[0;1] \times [0;1]$. Therefore the surface has the pinch point $P = \Phi(E)$.

We calculate the points of $\mathcal{C}(t,u)$ on the borders, i.e. the intersections of $\mathcal{C}(t,u)$ and of the lines $(u = 0)$, $(u = 1)$, $(t = 0)$, $(t = 1)$ in the parameter space. We obtain 4 points: ($t_1 \approx 0.7002714999$, $u_1 = 0$), ($t_2 \approx 0.6235730217$, $u_2 = 0$), ($t_3 = 1$, $u_3 \approx 0.4402786871$), ($t_4 = 1$, $u_4 \approx 0.8820099327$).

Now, we look for the points on $\mathcal{C}(t,u)$ in the box $[0;1] \times [0;1]$ that correspond to 4 points (t_i, u_i), $i = 1, \ldots, 4$ in order to detect the segment of $\mathcal{C}(t,u)$ corresponding to the self-intersections of the surface. (Two points on $\mathcal{C}(t,u)$ are called corresponding if their images by Φ are coincident on the singular locus of the surface. We also note that two points in the parameter space satisfying this condition lie on $\mathcal{C}(t,u)$. We see that a critical point does not have any corresponding point except itself). Hence, for $i = 1, \ldots, 4$, to find a point $(t,u) \in [0;1]^2$ corresponding to the point (t_i, u_i), we resolve the system $\Phi(t,u) = \Phi(t_i, u_i)$ in variables (t,u) in $[0;1]^2$. We obtain only the point ($t \approx 0.6632643380$, $u \approx 0.1700120872$) which correspond to the point (t_4, u_4). We denote these two points by B and A.

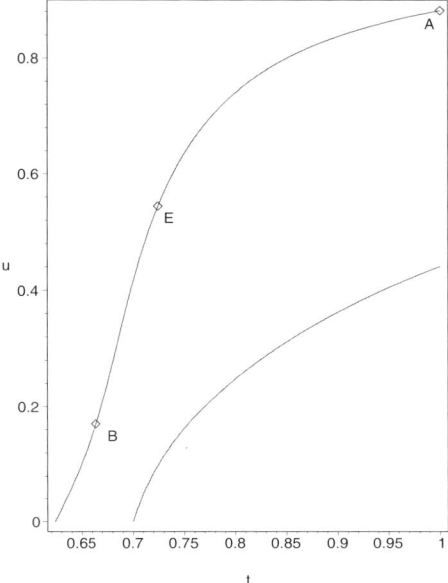

Fig. 6.2. The double locus $\mathcal{C}(t, u)$ in the space of parameters.

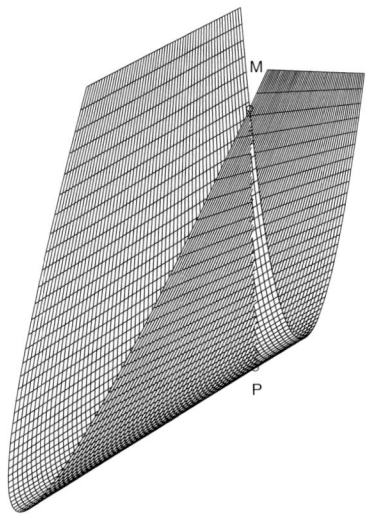

Fig. 6.3. The surface $\mathcal{S}$ with $(t = 0.6632643380..1, u = 0.1..0.9)$.

We have that the critical point E is on the segment (AB) of the curve $\mathcal{C}(t, u)$ and $\Phi(A) = \Phi(B) = M$. The images of the segments (BE) and (AE) of $\mathcal{C}(t, u)$ by Φ

are coincident. It is the self-intersection segment (MP) of the surface in

$$[0.6632643380; 1] \times [0.1; 0.9]$$

(it is the same in $[0; 1] \times [0; 1]$). The others segments in the figure (2) are phantom curves: they correspond to double points of the parameterization $\Phi(t_1, u_1) = \Phi(t_2, u_2)$ with $(t_1, u_1) \in [0; 1]^2$ but $(t_2, u_2) \notin [0; 1]^2$.

6.7 Conclusion

In this paper, we completed the classification of parametric surfaces of bidegree (1,2) over the complex field and over the real field. In a future work,we will also provide some results for the inverse problem: given a candidate (e.g. a segment of a line or of twisted cubic curve), we look for a patch (1,2) which includes this candidate as a subset of its singular locus. For instance we will characterize the ruled surfaces containing a twisted cubic curve and such that all generating lines cut twice the cubic curve, which are indeed parametric surfaces of bidegree $(1, 2)$.

Acknowledgements

We would like to thank Ragni Piene and the anonymous referees for their comments and suggestions. We acknowledge the partial support of the European Projects GAIA II (IST-2001-35512) and of the Network of Excellence Aim@Shape (IST NoE 506766).

References

1. L. Andersson, J. Peters, and N. Stewart, *Self-intersection of composite curves and surfaces*, Computer Aided Geometric Design, 15 (1998), pp. 507–527.
2. L. Busé and C. D'Andrea, *Inversion of parametrized hypersurfaces by means of subresultants*, preprint, (2004).
3. A. Cayley, *A second memoir on skew surfaces, otherwise scrolls*, Papers, vol 5, (1864), pp. 214–219.
4. A. Cayley, *A third memoir on skew surfaces, otherwise scrolls*, Papers, vol 6, (1868), pp. 312–328.
5. E.-W. Chionh and R. N. Goldman, *Degree, multiplicity, and inversion formulas for rational surfaces using u-resultants*, Computer Aided Geometric Design, 9 (1992), pp. 93–108.
6. A. Coffman, A. J. Schwartz, and C. Stanton, *The algebra and geometry of Steiner and other quadratically parametrizable surfaces*, Comput. Aided Geom. Des., 13 (1996), pp. 257–286.
7. L. Cremona, *Sulle superficie gobbe di quarto grado*, Memorie dell'Accademia delle Scienze dell'Istituto di Bologna, serie II, tomo 8 (1868), pp. 235–250; Opere, 2, 420.
8. T. Dokken, *Aspects of Intersection Algorithms and Approximation*, Thesis for the doctor philosophias degree, University of Oslo, Norway, 1997.

9. W.L.F. Degen, *The types of rational (2,1)-Bézier surfaces*, Computer Aided Geometric Design, 16(7), 1999, pp. 639–648.

10. T. Dokken, *Approximate implicitization*, Mathematical Methods for Curves and Surfaces, Oslo 200, T. Lyche and L.L. Schumaker (eds.), Vanderbilt University Press, 2001, pp. 81-102.

11. T. Dokken and J.B. Thomassen, *Overview of Approximate Implicitization*, in Topics in Algebraic Geometry and Geometric modeling, ed. Ron Goldman and Rimvydas Krasauskas, AMS series on Contemporary Mathematics CONM 334, 2003, pp 169-184.

12. W. L. Edge, *The theory of ruled surfaces*, Cambridge at the University Press, 1931.

13. M. Elkadi, T. Le, A. Galligo, *Parametrized surfaces in $\mathbb{P}^3$ of bidegree $(1,2)$*, in International Symposium on Symbolic and Algebraic Computation, J. Gutierre, ed., ACM, (2004), pp. 141–148.

14. M. Elkadi, B. Mourrain, *Residue and Implicitization Problem for Rational Surfaces*, Appl. Algebra Eng. Commun. Comput., 14 (2004), pp. 361–379.

15. G. Farin, *Curves and surfaces for computer aided geometric design. A practical guide*, Academic Press, Inc., Boston, MA, 1993.

16. A. Galligo and J. Pavone, *A sampling algorithm for parametric surface self-intersection*, preprint, (2004).

17. A. Galligo and M. Stillman, *The geometry of bicubic surfaces and splines*, preprint, (2004).

18. R. Hartshorne, *Algebraic Geometry*, Springer-Verlag, 1977.

19. E. Müller and J. L. Krames, *Konstruktive Behandlung der Regelflächen*, Vorlesur gen über Darstellende Geometrie, volume III, Franz Deuticke, Leipzig und Wien, 1931.

20. J-P. Pavone, *Auto-intersection des surfaces paramétrées réelles*, Thèse d'informatique de l'Université de Nice Sophia- Antipolis, Décembre 2004.

21. S. Perez-Diaz, J. Schicho, and J. Sendra, *Properness and inversion of rational parameterizations of surfaces*, Appl. Alg. Eng. Comm. Comp., 13 (2002), pp. 29–51.

22. R. Piene, *Singularities of some projective rational surfaces*, Computational methods for Algebraic Spline Surfaces (eds. T. Dokken, B. Jüttler), Springer-Verlag 2004, pp. 171–182.

23. R. Piene, G. Sacchiero, *Duality for rational normal scrolls*, Comm. Algebra 12 (9–10), (1984), pp. 1041–1066.

24. G. D. Reis, B. Mourrain, and J.-P. Técourt, *On the representations of 3d surfaces*, ECG Report, (2002).

25. G. Salmon, *Traité de Géométrie analytique à trois dimensions*, Paris, Gauthier-Villars, 1ère partie (1882), 2ème partie (1903), 3ème partie (1892).

26. I. Shafarevich, *Basic Algebraic Geometry*, New-York, Springer-Verlag, 1974.

27. C. Segre, *Recherches générales sur les courbes et les surfaces réglées algébriques (1ère partie courbes algébriques)*, Math. Ann., 30, 1887, pp. 203–226.

28. C. Segre, *Recherches générales sur les courbes et les surfaces réglées algébriques (2ème partie surfaces réglées algébriques)*, Math. Ann., 34, 1889, pp. 1–25.

29. J.B. Thomassen, *Self-Intersection Problems and Approximate Implicitization*, in Computational Methods for Algebraic Spline Surfaces, Springer, 2005, pp 155–170.

30. S. Zube, *Bidegree (2,1) parameterizable surfaces in projective 3-space*, Lithuanian Mathematical Journal, 1998, 38-3, pp 379–402.

31. S. Zube, *Correspondences and Geometry of (2,1)-Bézier surfaces*, Lithuanian Mathematical Journal, 2003, 43-1, pp 99–122.

Part III

Algorithms for geometric computing

In this part we collect six chapters which describe algorithms for geometric computing with curves and surfaces.

Beck and Schicho discuss the parameterization of planar rational curves over optimal field extensions, by exploiting the Newton polygon. Their method generates a parameterization in a field extension of degree one or two.

Ridges and umbilics of surfaces are among the objects studied in classical differential geometry, and they are of some interest for characterizing and analyzing the shape of a surface. In the case of polynomial parametric surfaces, these special curves are studied in the chapter by Cazals, Faugère, Pouget and Rouillier. In particular, the authors describe an algorithm which generates a certified approximation of the ridges. In order to illustrate the efficiency, the authors report on experiments where the algorithm is applied to Bézier surface patches.

Chau, Oberneder, Galligo and Jüttler report on several symbolic-numeric techniques for analyzing and computing the intersections and self-intersections of biquadratic tensor product Bézier surface patches. In particular, they explore how far one can go by solely using techniques from symbolic computing, in order to avoid potential robustness problems.

Cube decompositions by eigenvectors of quadratic multivariate splines are analyzed by Ivrissimtzis and Seidel. The results are related to subdivision algorithms, such as the tensor extension of the Doo–Sabin subdivision scheme.

A subdivision method for analyzing the topology of implicitly defined curves in two- and three-dimensional space are studied by Liang, Mourrain and Pavone. The method produces a graph which is isotopic to the curve. The authors also report on implementation aspects and on experiments with planar curves, such as ridge curves or self intersection curves of parameterized surfaces, and on silhouette curves of implicitly defined surfaces.

The final chapter of this volume, by Shalaby and Jüttler, describes techniques for the approximate implicitization of space curves and of surfaces of revolution. Both problems can be reduced to the planar situation. Special attention is paid to the problem of unwanted branches and singular points in the region of interest.

7

Curve Parametrization over Optimal Field Extensions Exploiting the Newton Polygon

Tobias Beck and Josef Schicho

Johann Radon Institute for Computational and Applied Mathematics,
Austrian Academy of Sciences
Tobias.Beck@oeaw.ac.at
Josef.Schicho@oeaw.ac.at

Summary. This paper describes an algorithm for rational parametrization of plane algebraic curves of genus zero. It exploits the shape of the Newton polygon. The computed parametrization has coefficients in an optimal field extension, which is of degree one or two.

7.1 Introduction

Given a bivariate polynomial $f \in \mathbb{K}[x, y]$ over a perfect field $\mathbb{K}$ (in particular any field of characteristic zero) we will describe a method to find a proper parametrization of the curve defined implicitly by f if it exists. That is we try to find $(X(t), Y(t)) \in \mathbb{L}(t)^2$ with $\mathbb{L} \mid \mathbb{K}$ an algebraic field extension such that $f(X(t), Y(t)) = 0$ and $(X(t), Y(t))$ induces a birational map from the affine line to the curve. One condition for the existence of such a parametrization is that f is absolutely irreducible, i.e. cannot be factored over any algebraic field extension of $\mathbb{K}$. In case f is absolutely irreducible, existence can be decided by computing a numerical invariant of the curve, namely its genus. This is the problem of finding a rational parametrization, a well-studied subject in algebraic geometry. There are already several algorithms, e.g. [10, 13, 14] and [18].

The complexity of the former algorithms is very sensitive with respect to the total degree of f. If f is sparse then one can take advantage by exploiting the shape of its Newton polygon (see remark 3). The algorithm described in [1] is the first to do so. The main idea there is to adapt an algorithm in [14] for curves in the projective plane to curves in a toric surface defined by the Newton polygon. In [1] and [14], the computed parametrizations have coefficients in a field extension of possibly large degree. On the other hand it is well-known that field extensions of degree at most 2 always suffice, and there are algorithms that compute parametrizations using an optimal field extension, e.g. [15] and [18].

The original result in this paper is an adaptation of the method described in [18] to the toric case. This gives a parametrization algorithm producing a parametrization in an optimal field extension which exploits the shape of the Newton polygon. We tried to make the explanation of the necessary concepts and the description of the algorithm as self-contained as possible.

Once and for all let $\mathbb{K}$ denote a perfect field, the field of definition, and $\overline{\mathbb{K}}$ an algebraic closure of $\mathbb{K}$. Further let $f \in \mathbb{K}[x, y]$ be an absolutely irreducible polynomial, i.e. irreducible in $\overline{\mathbb{K}}[x, y]$.

7.2 Toric geometry

In this section we introduce just as much of toric geometry as we need in this paper. A good general introduction to toric geometry is [3]. We focus on the fact that smooth toric surfaces are generalizations of the standard surfaces $\mathbb{P}^2_{\mathbb{K}}$ and $\mathbb{P}^1_{\mathbb{K}} \times \mathbb{P}^1_{\mathbb{K}}$.

7.2.1 The projective plane $\mathbb{P}^2_{\mathbb{K}}$

The projective plane is the set of points $(\bar{v} : \bar{x} : \bar{y})$ subject to the equivalence relation $(\bar{v} : \bar{x} : \bar{y}) = (\lambda \bar{v} : \lambda \bar{x} : \lambda \bar{y})$ for $\lambda \neq 0$. It can be covered by 3 affine planes, which are open subsets, depending on whether $\bar{v} \neq 0$, $\bar{x} \neq 0$ or $\bar{y} \neq 0$. We can introduce *local coordinates* on each of these open subsets:

$$(\bar{v} : \bar{x} : \bar{y}) = \begin{cases} (1 : \frac{\bar{x}}{\bar{v}} : \frac{\bar{y}}{\bar{v}}) =: (1 : u_1 : v_1) & \text{if } \bar{v} \neq 0 \\ (\frac{\bar{v}}{\bar{x}} : 1 : \frac{\bar{y}}{\bar{x}}) =: (v_2 : 1 : u_2) & \text{if } \bar{x} \neq 0 \\ (\frac{\bar{v}}{\bar{y}} : \frac{\bar{x}}{\bar{y}} : 1) =: (u_3 : v_3 : 1) & \text{if } \bar{y} \neq 0 \end{cases}$$

If both sides are defined, i.e. on the intersection of open subsets, we see that

$$v_i = u_{i-1}^{-1}, \ u_i = v_{i-1} u_{i-1}^{-1}. \tag{7.1}$$

Here we assumed for convenience that indices are cyclically arranged, i.e. $u_3 = u_0$ and $v_3 = v_0$.

The transformation rules for the local coordinates can also be described using a lattice polygon: Draw an isosceles triangle with vertices in $\mathbb{Z}^2$ as in figure 7.1 left, label the vertices cyclically from 1 to 3 and attach two minimal direction vectors $\mathbf{u}_i$ and $\mathbf{v}_i$ to each vertex. Then we find the relations

$$\mathbf{v}_i = (-\mathbf{u}_{i-1}), \ \mathbf{u}_i = \mathbf{v}_{i-1} + (-\mathbf{u}_{i-1})$$

which correspond to (7.1) when passing from additive to multiplicative writing.

7.2.2 The ruled surface $\mathbb{P}^1_{\mathbb{K}} \times \mathbb{P}^1_{\mathbb{K}}$

The previous example is no coincidence. A similar analogy holds in case of the ruled surface $\mathbb{P}^1_{\mathbb{K}} \times \mathbb{P}^1_{\mathbb{K}}$ (if a little care is taken when numbering affine charts).

$\mathbb{P}^1_{\mathbb{K}} \times \mathbb{P}^1_{\mathbb{K}}$ can be seen as the set of points $(\bar{u}, \bar{v}, \bar{x}, \bar{y})$ subject to the equivalence relation $(\bar{u}, \bar{v}, \bar{x}, \bar{y}) = (\lambda \bar{u}, \mu \bar{v}, \lambda \bar{x}, \mu \bar{y})$ for $\lambda \mu \neq 0$. It can be covered by 4 affine planes, which are open subsets, depending on whether $\bar{u}\bar{v} \neq 0$, $\bar{x}\bar{v} \neq 0$, $\bar{x}\bar{y} \neq 0$ or $\bar{u}\bar{y} \neq 0$. Again we introduce local coordinates:

$$(\bar{u}, \bar{v}, \bar{x}, \bar{y}) = \begin{cases} (1, 1, \frac{\bar{x}}{\bar{u}}, \frac{\bar{y}}{\bar{v}}) =: (1, 1, u_1, v_1) & \text{if } \bar{u}\bar{v} \neq 0 \\ (\frac{\bar{u}}{\bar{x}}, 1, 1, \frac{\bar{y}}{\bar{v}}) =: (v_2, 1, 1, u_2) & \text{if } \bar{x}\bar{v} \neq 0 \\ (\frac{\bar{u}}{\bar{x}}, \frac{\bar{v}}{\bar{y}}, 1, 1) =: (u_3, v_3, 1, 1) & \text{if } \bar{x}\bar{y} \neq 0 \\ (1, \frac{\bar{v}}{\bar{y}}, \frac{\bar{x}}{\bar{u}}, 1) =: (1, u_4, v_4, 1) & \text{if } \bar{u}\bar{y} \neq 0 \end{cases}$$

Now changing from one coordinate system to the other we find

$$v_i = u_{i-1}^{-1}, \; u_i = v_{i-1}$$

and the coordinate change could be derived from a rectangle (see figure 7.1 right):

$$\mathbf{v}_i = (-\mathbf{u}_{i-1}), \; \mathbf{u}_i = \mathbf{v}_{i-1}$$

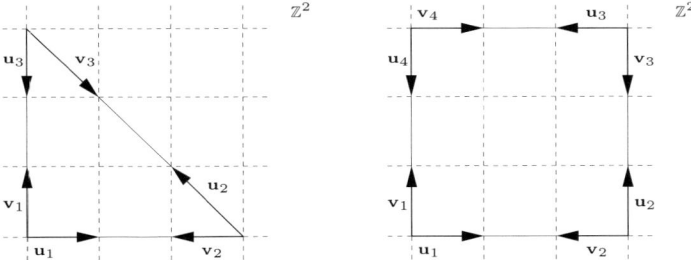

Fig. 7.1. Isosceles triangles and squares

7.2.3 Smooth toric surfaces

The preceding two examples give rise to a general construction.

Smooth polygons

Let $\Pi \subset \mathbb{R}^2$ be a *convex lattice polygon*, that is a convex polygon whose vertices have integral coordinates. Label its vertices cyclically and attach two minimal direction vectors $\mathbf{u}_i$ and $\mathbf{v}_i$ to each vertex.

To proceed as in the examples, we would need that each pair $(\mathbf{u}_i, \mathbf{v}_i)$ can be expressed as a $\mathbb{Z}$-linear combination using any other pair $(\mathbf{u}_j, \mathbf{v}_j)$. For this it is sufficient that each of the pairs generates the entire integer lattice, i.e.

$$\mathbb{Z}\mathbf{u}_i + \mathbb{Z}\mathbf{v}_i = \mathbb{Z}^2. \tag{7.2}$$

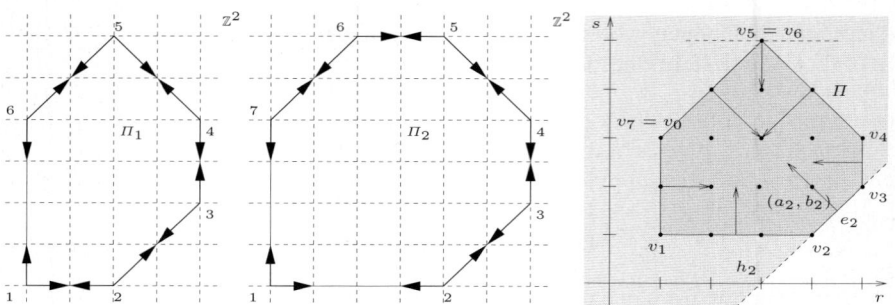

Fig. 7.2. General convex polygons

If the polygon has this shape we say it is *smooth* and an abstract smooth surface can be constructed in pretty much the same way as above.

A general convex polygon will unfortunately not fulfill condition (7.2), see for example Π_1 in figure 7.2. Here the pair of vectors attached to the edge 5 does not span the entire lattice. However the situation can be fixed, when we consider instead the polygon Π_2, which was generated from Π_1 by "inserting" an additional edge.

To describe the following procedure precisely it is convenient to work with inner normal vectors instead of edge direction vectors and introduce some further terminology, see also figure 7.2 right. Let $\Pi \subset \mathbb{R}^2$ be the convex hull of a finite set of lattice points in $\mathbb{Z}^2$.

For any pair of relatively prime integers a, b, let $c(a, b) \in \mathbb{Z}$ be the minimal value of $ar + bs$, where $(r, s) \in \Pi$. Then Π is a finite intersection of say n support half planes

$$h_i := \{(r, s) \in \mathbb{R}^2 \mid a_i r + b_i s \ge c_i\}$$

where the (a_i, b_i) are inward pointing normal vectors and $c_i := c(a_i, b_i)$. We assume them to be cyclically arranged, i.e. $a_{i-1} b_i - a_i b_{i-1} > 0$ (setting $a_0 := a_n$ and $b_0 := b_n$). We also give names to the edges and the vertices of intersection

$$e_i := \{(r, s) \in \Pi \mid a_i r + b_i s = c_i\} \text{ and } v_i := e_i \cap e_{i-1}.$$

Note that the set of half planes is not uniquely defined, there may be redundant half planes where an edge meets Π in one vertex (in this case, some of the vertices v_i will coincide). Using the redundancy of this representation, we can enforce a condition similar to (7.2), namely we may assume

$$a_{i-1} b_i - a_i b_{i-1} = 1 \text{ for } 1 \le i \le n. \tag{7.3}$$

This condition holds for example for the polygon in figure 7.2 right because an additional normal vector was introduced at the vertex $v_5 = v_6$. If only two edges meet in one vertex conditions (7.2) and (7.3) are easily shown to be the same. In the following we sketch an algorithm to ensure condition (7.3). This makes it more precise what "inserting an edge" meant for II_1 in figure 7.2. It basically corresponds to the resolution of a toric surface, see e.g. [4].

The values $\det((a_{i-1}, b_{i-1})^T, (a_i, b_i)^T) = a_{i-1}b_i - a_i b_{i-1}$ are invariant under unimodular transformations (i.e. linear transformations of the vectors by an integral matrix with determinant 1). Assume that $a_{i_0-1}b_{i_0} - a_{i_0}b_{i_0-1} > 1$ for some i_0. By a suitable unimodular transformation we may assume $(a_{i_0}, b_{i_0}) = (0, 1)$. It follows that $a_{i_0-1} > 1$.

We insert a new index, for simplicity say $i_0 - \frac{1}{2}$, set $a_{i_0-\frac{1}{2}} := 1$ and determine $b_{i_0-\frac{1}{2}}$ by integer division s.t. $0 \le a_{i_0-1}b_{i_0-\frac{1}{2}} - b_{i_0-1} < a_{i_0-1}$. It follows

$$a_{i_0-\frac{1}{2}}b_{i_0} - a_{i_0}b_{i_0-\frac{1}{2}} = 1 \cdot 1 - 0 \cdot b_{i_0-\frac{1}{2}} \qquad = 1 \qquad \text{and}$$

$$a_{i_0-1}b_{i_0-\frac{1}{2}} - a_{i_0-\frac{1}{2}}b_{i_0-1} = a_{i_0-1}b_{i_0-\frac{1}{2}} - 1 \cdot b_{i_0-1} < a_{i_0-1}$$

$$= a_{i_0-1} \cdot 1 - 0 \cdot b_{i_0-1} \qquad = a_{i_0-1}b_{i_0} - a_{i_0}b_{i_0-1}.$$

By inserting the additional support half plane with normal vector $(a_{i_0-\frac{1}{2}}, b_{i_0-\frac{1}{2}})$ and support line through the vertex v_{i_0}, we "substitute" the value $a_{i_0-1}b_{i_0} - a_{i_0}b_{i_0-1}$ by the smaller value $a_{i_0-1}b_{i_0-\frac{1}{2}} - a_{i_0-\frac{1}{2}}b_{i_0-1}$ and add $a_{i_0-\frac{1}{2}}b_{i_0} - a_{i_0}b_{i_0-\frac{1}{2}} = 1$ to the list. All other values stay fixed. Repeating this process statement (7.3) can be achieved.

Constructing the surface

Now we construct a surface following the examples of $\mathbb{P}^2_{\mathbb{K}}$ and $\mathbb{P}^1_{\mathbb{K}} \times \mathbb{P}^1_{\mathbb{K}}$. For $1 \le i \le n$ let $U_i := \mathbb{A}^2_{\mathbb{K}}$ be copies of the affine plane with coordinates u_i and v_i. Again we identify U_0 and U_n. We denote the coordinate axes by $L_i := \{(u_i, v_i) \in U_i \mid u_i = 0\}$ and $R_i := \{(u_i, v_i) \in U_i \mid v_i = 0\}$ and define open embeddings of the algebraic torus $T := (\overline{\mathbb{K}}^*)^2$ where $\overline{\mathbb{K}}^* = \overline{\mathbb{K}} \setminus \{0\}$:

$$\psi_i : T \to U_i : (x, y) \mapsto (u_i, v_i) = (x^{b_i}y^{-a_i}, x^{-b_{i-1}}y^{a_{i-1}})$$

The isomorphic image of ψ_i is $U_i \setminus (L_i \cup R_i)$ and there it has the inverse

$$U_i \setminus (L_i \cup R_i) \to T : (u_i, v_i) \mapsto (x, y) = (u_i^{a_{i-1}}v_i^{a_i}, u_i^{b_{i-1}}v_i^{b_i}).$$

For $i, j \in \{1, \ldots, n\}, i \ne j$ we define open subsets

$$U_{i,j} := \begin{cases} U_i \setminus L_i & \text{if } i \equiv j - 1 \bmod n, \\ U_i \setminus R_i & \text{if } i \equiv j + 1 \bmod n, \\ U_i \setminus (L_i \cup R_i) = \psi_i(T) & \text{else.} \end{cases}$$

For $1 \le i \le n$ the following maps are mutually inverse and therefore isomorphisms:

$$\varphi_{i-1,i} : U_{i-1,i} \to U_{i,i-1} : (u_{i-1}, v_{i-1}) \mapsto (u_i, v_i) = (u_{i-1}^{a_i-2b_i-a_ib_i-2}v_{i-1}, u_{i-1}^{-1})$$

$$\varphi_{i,i-1} : U_{i,i-1} \to U_{i-1,i} : (u_i, v_i) \mapsto (u_{i-1}, v_{i-1}) = (v_i^{-1}, u_iv_i^{a_i-2b_i-a_ib_i-2})$$

If i and j are non-neighboring indices we set

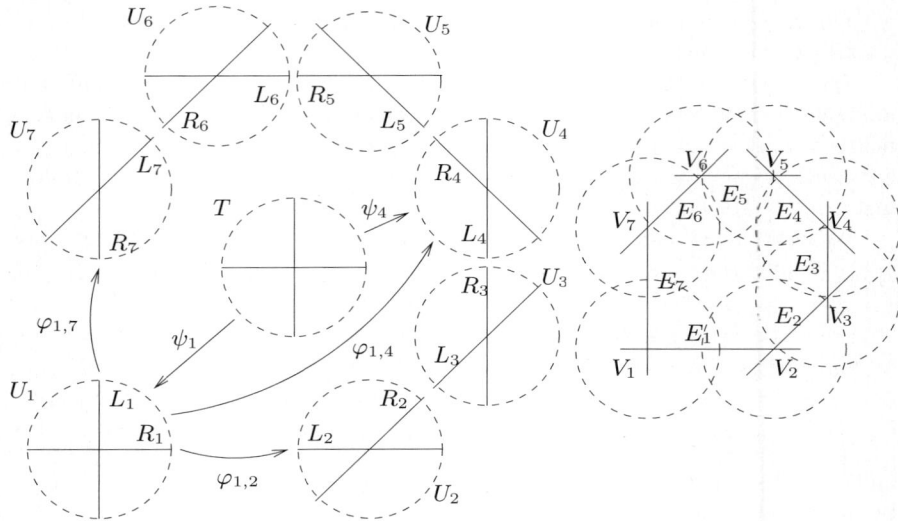

Fig. 7.3. Construction of the toric surface S

$$\varphi_{i,j} := \psi_j \circ \psi_i^{-1} : U_{i,j} \to U_{j,i}.$$

The above morphisms are compatible in the sense that $\varphi_{j,i} = \varphi_{i,j}^{-1}$ and for each k we have $\varphi_{i,j}(U_{i,j} \cap U_{i,k}) = U_{j,i} \cap U_{j,k}$ and $\varphi_{i,k} = \varphi_{j,k} \circ \varphi_{i,j}$.

Now we can define an *abstract surface* S as a finite quotient $\overline{S}/\equiv$ where $\overline{S} := \dot{\bigcup}_i U_i$ is the disjoint union of affine planes. We identify points along the morphisms $\varphi_{i,j}$, see figure 7.3. That is for $i \neq j$ a point $(a,b) \in U_i$ and a point $(c,d) \in U_j$ are equivalent $(a,b) \equiv (c,d)$ if $(c,d) = \varphi_{i,j}(a,b)$ whenever this expression is defined.

This corresponds to the general *gluing construction* for schemes, see e.g. [9, exercise II.2.12]. By abuse of notation we will from now on identify U_i and its image in S. Then $\{U_i\}_{1 \leq i \leq n}$ is an open cover.

It is not hard to see that $E_i := R_i \cup L_{i+1}$ is an irreducible curve on S and isomorphic to $\mathbb{P}^1_{\overline{\mathbb{K}}}$. We call it an *edge curve*. The curves E_{i-1} and E_i intersect transversally in a point $V_i \in U_i$, corresponding to the origin $(u_i, v_i) = (0,0)$ of the corresponding chart. For non-neighboring indices i and j the edge curves E_i and E_j are disjoint. The complement of the union of all edge curves is the torus T, which is also the intersection of all open sets U_i.

Comparing figures 7.2 and 7.3 suggests that there are certain correspondences between a smooth polygon, here Π_2, and the toric surface constructed from it. For example there is a line $\mathbb{P}^1_{\overline{\mathbb{K}}}$ corresponding to each edge of the polygon and they intersect accordingly. On the other hand the constructed surface is invariant with respect to the scaling of the polygon and even the actual length of its edges. The only important data is the set of normal vectors.

We briefly summarize important properties of the toric surface S:

- S is smooth, in fact it is covered by affine planes. Hence we are locally working with polynomials.
- S is a *complete* algebraic variety (see for example [3] or [5]).
- S contains the torus $T = (\overline{\mathbb{K}}^*)^2$ as a dense open subset.

7.2.4 Completion of the curve

The parametrization problem for curves is a problem of birational geometry. To solve it, we have to apply certain theorems of "global content". Therefore we have to study a complete model of our curve, that is a curve without any missing points.

The Newton polygon $\Pi(f) \subset \mathbb{R}^2$ is defined as the convex hull of all lattice points $(r, s) \in \mathrm{Supp}(f)$ (i.e. all $(r, s) \in \mathbb{Z}^2$ s.t. $x^r y^s$ appears with a non-zero coefficient in f). An absolutely irreducible polynomial $f \in \mathbb{K}[x, y]$ defines an irreducible curve in the affine plane $\mathbb{A}^2_{\overline{\mathbb{K}}}$. If $\Pi(f)$ is *non-degenerate*, i.e. has dimension 2, then f also defines a curve on the torus $T \subset \mathbb{A}^2_{\overline{\mathbb{K}}}$.

From now on we also fix the surface S which is constructed from $\Pi(f)$ as in the previous section. S contains the torus and hence we can define C to be the Zariski closure of the curve defined by f on the torus. If the half planes used for the construction of S are determined by the integers a_i, b_i, c_i then C is defined by the polynomials

$$f_i(u_i, v_i) := u_i^{-c_i-1} v_i^{-c_i} f(u_i^{a_i-1} v_i^{a_i}, u_i^{b_i-1} v_i^{b_i}) \tag{7.4}$$

within the open subsets $U_i \subset S$. As a closed subset of a complete space, C is *complete* itself.

For example if f is a dense polynomial with respect to total degree, meaning that it contains all monomials up to a certain degree, then $\Pi(f)$ is a triangle. If f is a dense polynomial with respect to bidegree, then $\Pi(f)$ is a rectangle. So in the first case we would work inside $\mathbb{P}^2_{\overline{\mathbb{K}}}$, in the second case inside $\mathbb{P}^1_{\overline{\mathbb{K}}} \times \mathbb{P}^1_{\overline{\mathbb{K}}}$. In general the surface is adapted to the Newton polygon, which is of course a much finer shape parameter than any notion of degree.

We consider a polynomial to be sparse, if the shape of its Newton polygon differs from an isosceles triangle. In this case our algorithm is more efficient than algorithms relying on a projective embedding.

Throughout this article we will *always* implicitly assume that f is absolutely irreducible and $\Pi(f)$ is non-degenerate. For parametrizing in the other cases it is easy to devise specialized algorithm, see also [1].

7.3 Divisors

In this section we introduce divisors and linear systems associated to them.

7.3.1 General and $\mathbb{K}$-rational divisors

Let X be a smooth, irreducible $\overline{\mathbb{K}}$-variety with field of definition $\mathbb{K}$. By this we mean that X is locally defined by equations with coefficients in $\mathbb{K}$. Let $\overline{\mathbb{K}}(X)$ denote the function field of X. Then the Galois group $\mathrm{Gal}(\overline{\mathbb{K}} \mid \mathbb{K})$ acts on $\overline{\mathbb{K}}(X)$. For any field $\mathbb{L}$ s.t. $\mathbb{K} \subset \mathbb{L} \subset \overline{\mathbb{K}}$ we define the restricted function field $\mathbb{L}(X) := \{g \in \overline{\mathbb{K}}(X) \mid \sigma(g) = g$ for all $\sigma \in \mathrm{Gal}(\overline{\mathbb{K}} \mid \mathbb{L})\}$.

Irreducible closed $\overline{\mathbb{K}}$-subvarieties of X of codimension 1 are also called *prime (Weil) divisors*. A *general divisor* D is defined to be a finite formal sum $D = \sum_i n_i D_i$ with $n_i \in \mathbb{Z}$ and D_i prime divisors. For curves one defines the *degree* of the divisor as $\deg(D) = \sum_i n_i$. The set of divisors thus forms a free Abelian group $\mathrm{Div}(X)$. An *effective divisor* is a non-negative linear combination of prime divisors.

One associates to a nonzero rational function $g \in \overline{\mathbb{K}}(X)$ its *principal divisor* (g). Roughly speaking g has poles and zeroes with certain multiplicities on X along subvarieties of codimension 1; then (g) is the divisor of zeroes minus the divisor of poles (with multiplicities). We say that two divisors are *equivalent* if and only if their difference is principal. For a more precise elaboration see [16, III.1.1].

Let $D \in \mathrm{Div}(X)$. The Galois group $\mathrm{Gal}(\overline{\mathbb{K}} \mid \mathbb{K})$ also acts on the divisors $\mathrm{Div}(X)$. For any field extension $\mathbb{L}$ s.t. $\mathbb{K} \subset \mathbb{L} \subset \overline{\mathbb{K}}$ we say that D is an $\mathbb{L}$-*rational divisor* if and only if it is invariant under the Galois group $\mathrm{Gal}(\overline{\mathbb{K}} \mid \mathbb{L})$. For example if $g \in \mathbb{L}(X)$ then the principal divisor (g) is an $\mathbb{L}$-rational divisor. From the definition it follows that also $\mathrm{Gal}(\mathbb{L} \mid \mathbb{K})$ acts on the set of $\mathbb{L}$-rational divisors if $\mathbb{L} \mid \mathbb{K}$ is itself Galois.

The *linear system* $\mathcal{L}_X(D)$ of rational functions on X associated to a divisor D is the $\overline{\mathbb{K}}$-vector space of rational functions $g \in \overline{\mathbb{K}}(X)$ s.t. $D + (g)$ is effective. $\mathbb{K}$-rational divisors are of particular interest because the corresponding linear systems can be represented without introducing field extensions.

Lemma 1. *Let X be a smooth, projective $\overline{\mathbb{K}}$-variety with field of definition $\mathbb{K}$. If D is a $\mathbb{K}$-rational divisor then $\mathcal{L}_X(D)$ has a basis in $\mathbb{K}(X)$ (or $\mathcal{L}_X(D) = \emptyset$).*

Proof. Since X is projective the vector space $\mathcal{L}_X(D)$ is finite-dimensional (or empty, see [9, Theorem II.5.19]). Therefore we can assume without loss of generality that $\langle b_1, \ldots, b_m \rangle_{\overline{\mathbb{K}}} = \mathcal{L}_X(D)$ with $b_i \in \mathbb{L}(X)$ for some Galois extension $\mathbb{L} \mid \mathbb{K}$. Set $V := \langle b_1, \ldots, b_m \rangle_{\mathbb{L}} = \mathbb{L}(X) \cap \mathcal{L}_X(D)$.

Let $\sigma \in \mathrm{Gal}(\mathbb{L} \mid \mathbb{K})$ and assume that $g \in V$, i.e. $g \in \mathbb{L}(X)$ and $(g) \geq -D$. Then $(\sigma g) = \sigma(g) \geq -\sigma D = -D$, because D is $\mathbb{K}$-rational. In other words $\sigma g \in V$ and the Galois group $\mathrm{Gal}(\mathbb{L} \mid \mathbb{K})$ acts semi-linearly on V. Let now $V_0 := \{g \in V \mid \sigma g = g$ for all $\sigma \in \mathrm{Gal}(\mathbb{L} \mid \mathbb{K})\}$.

Then by [12, Lemma 2.13.1] the canonical map $V_0 \otimes_{\mathbb{K}} \mathbb{L} \to V$ is an isomorphism. Since V_0 is fixed by $\mathrm{Gal}(\mathbb{L} \mid \mathbb{K})$ we have $V_0 \subset \mathbb{K}(X)$. Choose a basis of V_0. $\qquad$

7.3.2 Divisors on toric surfaces

For surfaces, prime divisors correspond to irreducible closed curves on the surface. If S is a toric surface as constructed in section 7.2 then it is locally isomorphic to

$\mathbb{A}^2_{\overline{\mathbb{K}}}$. In this case $g \in \overline{\mathbb{K}}(S)$ can be written as a rational function in $\overline{\mathbb{K}}(u_i, v_i)$ for all local coordinates u_i and v_i. The zeroes and poles (and their multiplicities) can be read easily from the (absolute) factorization of the reduced representation.

On a toric surface we have a set of distinguished prime divisors, namely the edge curves E_i, also called the toric invariant prime divisors. They are clearly $\mathbb{K}$-rational. The linear systems of toric invariant divisors can easily be described by support conditions.

Lemma 2. *Let* $D = \sum_{1 \leq i \leq n} -\tilde{c}_i E_i \in \mathrm{Div}(S)$ *and define the polygon* $\Pi :=$ $\bigcap_{1 \leq i \leq n}\{(r, s) \in \mathbb{R}^2 \mid a_i r + b_i s \geq \tilde{c}_i\}$. *Then* $\mathcal{L}_S(D) = \langle x^r y^s \mid (r, s) \in \Pi \cap \mathbb{Z}^2 \rangle_{\overline{\mathbb{K}}}$.

Proof. See [1, Corollary 8].

In particular $\mathcal{L}_S(D) \neq \emptyset$ if and only if $\Pi \neq \emptyset$ and the basis is obviously contained in $\mathbb{K}(S)$.

Remark 3. In our algorithm we use spaces of polynomials as above that are supported approximately on the Newton polygon of f. If we considered only the degree of the defining equation, the linear systems from above would correspond to spaces of polynomials supported approximately on an isosceles triangle containing the Newton polygon. As mentioned before, we consider a polynomial to be sparse, if its Newton polygon differs from such a triangle. Therefore for sparse polynomials these vector spaces become smaller and our algorithm becomes more efficient.

7.3.3 Divisors on smooth curves

For a smooth curve $\widetilde{C}$, prime divisors correspond to points on the curve. A rational function $g \in \overline{\mathbb{K}}(\widetilde{C})$ can be developed at any point $P \in \widetilde{C}$ as a Laurent series with respect to a local parameter.

More precisely let $P \in \widetilde{C}$ be a point. Then the local ring at the point P is regular and we can find an injective homomorphism $\mathcal{O}(\widetilde{C})_P \to \overline{\mathbb{K}}[[t]]$ to a power series ring. This homomorphism induces a homomorphism $\varphi_P : \overline{\mathbb{K}}(\widetilde{C}) \to \overline{\mathbb{K}}((t))$. We may assume that φ_P is *primitive*, i.e. $\mathrm{img}(\varphi_P) \not\subset \overline{\mathbb{K}}((t^d))$ for any $d > 1$. Then $\nu_P : \overline{\mathbb{K}}(\widetilde{C}) \to \mathbb{Z} : g \mapsto \mathrm{ord}_t(\varphi_P(h))$ is a *discrete valuation* of the function field $\overline{\mathbb{K}}(\widetilde{C})$ over $\overline{\mathbb{K}}$ with *center* P (see e.g. [19]).

Now if $\nu_P(g) > 0$ then g has a zero at P with multiplicity $\nu_P(g)$, if $\nu_P(g) < 0$ then g has a pole at P with multiplicity $-\nu_P(g)$.

7.3.4 Divisors on singular curves

If C is a complete, singular curve we may consider a *resolution of singularities* $\pi : \widetilde{C} \to C$. I.e. $\widetilde{C}$ is a complete, smooth curve and π is a regular, *birational map*. Such a resolution is known to exist and actually one can take the *normalization* of C (see [16, II.5.1]).

Up to isomorphism the resolution of a complete curve is unique (which follows from [16, II.4.4, Cor. 2]). We can effectively deal with the divisors $\mathrm{Div}(\widetilde{C})$ by identifying prime divisors on $\widetilde{C}$ with discrete valuations of $\overline{\mathbb{K}}(C)$ over $\overline{\mathbb{K}}$ (and $\overline{\mathbb{K}}(C)$ is isomorphic to $\overline{\mathbb{K}}(\widetilde{C})$ because of birationality). We never have to construct the curve $\widetilde{C}$ explicitly.

Above we have seen how a point $P \in \widetilde{C}$ induces a discrete valuation ν_P of $\overline{\mathbb{K}}(C)$. On the other hand let ν be such a discrete valuation. Since $\widetilde{C}$ is complete ν has a center $P_\nu \in \widetilde{C}$. This way $P \mapsto \nu_P$ and $\nu \mapsto P_\nu$ constitute a one-one correspondence between points of $\widetilde{C}$ and discrete valuations of $\overline{\mathbb{K}}(C)$.

Also C is complete and so we let $\mathrm{Center}(\nu)$ denote the center of a valuation ν in C. Note that with the above notation $\mathrm{Center}(\nu) = \pi(P_\nu)$. For any subset $M \subset C$ we define $\mathcal{V}_M$ to be the set of all discrete valuations ν of $\overline{\mathbb{K}}(C)$ s.t. $\mathrm{Center}(\nu) \in M$.

7.4 Rational curves

It is well-known that a curve is parametrizable if and only if it has *genus* zero. In this section we will show how to compute the genus in our setting. Afterwards we give the general idea of a curve parametrization algorithm.

7.4.1 The genus of a curve

To each point $Q \in C$ one can associate its *delta invariant* δ_Q. It is a measure of singularity, which is defined as the length of the quotient of the integral closure of the local ring by the local ring at Q (see [9, exercise IV.1.8]). For instance, if Q is an ordinary singularity of multiplicity μ, i.e. a self-intersection point where μ branches meet transversally, then $\delta_Q = \frac{\mu(\mu-1)}{2}$. In particular $\delta_Q = 0$ for Q smooth.

If $\Pi \subset \mathbb{R}^2$ is a lattice polygon we denote by $\#(\Pi) := |\Pi \cap \mathbb{Z}^2|$ the number of lattice points in Π. We also write Π° for the polygon spanned by the interior points. In the toric situation the genus can be computed as follows:

Theorem 4. *The genus of C is equal to the number of interior lattice points of $\Pi(f)$ minus the sum of the delta invariants of all points on C:*

$$\mathrm{genus}(C) = \#(\Pi(f)^\circ) - \sum_{Q \in C} \delta_Q$$

Proof. See [1, Proposition 9].

The sum actually ranges over the singular points of C only.

Remark 5. Assume that $\Pi(f)$ is an isosceles triangle with vertices $(0,0)$, $(n,0)$ and $(0,n)$ and that all singularities of the curve are ordinary. Then the number of interior points is equal to $\frac{(n-1)(n-2)}{2}$ and we recover the well-known genus formula for plain curves.

7.4.2 Parametrizing a curve

Now assume that C is an irreducible curve of genus$(C) = 0$ and $\pi : \widetilde{C} \to C$ a resolution. Assume that $D \in \text{Div}(\widetilde{C})$ is a divisor with $d := \deg(D) \geq 1$ and $\{s_0, \ldots, s_d\} \subset \mathcal{L}_{\widetilde{C}}(D)$ is a basis of the corresponding linear system. Using the isomorphism $\overline{\mathbb{K}}(C) \cong \overline{\mathbb{K}}(\widetilde{C})$ we can define a birational map

$$C \dashrightarrow \mathbb{P}^d_{\overline{\mathbb{K}}} : Q \mapsto [s_0(Q) : \cdots : s_d(Q)]$$

which sends C to a *rational normal curve* $C' \subseteq \mathbb{P}^d_{\overline{\mathbb{K}}}$.

Now if D is a $\mathbb{K}$-rational divisor then we can assume $\{s_0, \ldots, s_d\} \subset \mathbb{K}(C)$ by lemma 1. In other words the corresponding map and also its rational inverse do not require any field extension. The best thing would thus be to find a $\mathbb{K}$-rational divisor of degree 1 because that would result immediately in a parametrization of C without field extension. The existence of such a divisor however cannot be guaranteed and is equivalent to the existence of a rational point on C.

On the other hand the existence of a $\mathbb{K}$-rational *anticanonical divisor* is always guaranteed and in the case of a rational curve it has degree 2 (see section 7.5). The idea of the parametrization algorithms in [10] and [18] can therefore be explained as follows: Compute the linear system associated to a $\mathbb{K}$-rational anticanonical divisor. This system defines a birational map from C to a *conic* $C' \subset \mathbb{P}^2_{\mathbb{K}}$ (see 7.7.5 in the example section). The parametrization of C' is an easy task once a point on C' is known. The algebraic degree of the resulting parametrization obviously depends on the degree of the field extension needed to define that point. Hence it is at most two and the problem of parametrizing a rational curve using a minimal field extension is reduced to the problem of finding a rational point on a conic if it exists. This task is not straight forward. For example if $\mathbb{K} = \mathbb{Q}$ we refer to [11] and [17]. In this case rational points may also be found by the function `RationalPoint` of the computer algebra system Magma [2].

7.5 An anticanoncial divisor

We are in the situation $\widetilde{C} \xrightarrow{\pi} C \xrightarrow{\iota} S$ where C is an irreducible curve, embedded in the smooth toric surface S, and $\widetilde{C}$ is a resolution. Throughout this section we further assume that genus$(\widetilde{C}) = $ genus$(C) = 0$.

7.5.1 Support divisors

For $l, k \in \mathbb{Z}/n\mathbb{Z}$ we write $[l, k] = \{l, l+1, \ldots, k\} \subset \mathbb{Z}/n\mathbb{Z}$ for the set of cyclically consecutive indices between l and k. From now on I will always denote an "interval", i.e. $I = [l, k]$ or $I = \emptyset$, and δ_I will be its characteristic function, i.e. $\delta_I(i) = 1$ if $i \in I$ and $\delta_I(i) = 0$ else. Set

$$D_I := \sum_{i \in [1,n]} (-c_i - 1 + \delta_I(i)) E_i \in \text{Div}(S).$$

We call D_I a *support divisor* because the corresponding linear system $\mathcal{L}_S(D_I)$ on the surface can be described by simple support conditions. To this end let

$$\Pi_I := \bigcap_{i \in [1,n]} \{(r, s) \mid a_i r + b_i s \geq c_i + 1 - \delta_I(i)\}.$$

In other words Π_I is constructed as the convex hull of the lattice points in $\Pi(f)$ with edges e_i removed for $i \notin I$. In particular $\Pi_{[1,n]} = \Pi(f)$ and $\Pi_\emptyset = \Pi(f)^\circ$. With this notation we have $\mathcal{L}_S(D_I) = \langle x^r y^s \mid (r, s) \in \Pi_I \cap \mathbb{Z}^2 \rangle_{\overline{\mathbb{K}}}$ by lemma 2 (see also the Ansatz polynomials in 7.7.3 in the example section).

Note that since C is not a component of D_I the intersection divisor

$$\widetilde{D}_I := (\pi \circ \iota)^*(D_I) \in \mathrm{Div}(\widetilde{C})$$

is also well defined. This is sometimes called the *pullback* along $\pi \circ \iota$, see [16, III.1.2]. Further we can explicitly give its degree. The integer length of an edge of a lattice polygon is the number of lattice points on that edge minus 1. Let d_I denote the sum over the integer lengths of the edges i of $\Pi(f)$ for $i \in I$.

Lemma 6. *We have* $\deg(\widetilde{D}_I) = 2\#(\Pi(f)^\circ) + d_I - 2$.

Proof. See [1, Lemma 12]. ∎

7.5.2 Twists of principal divisors and valuations

A part of the principal divisor (h) of a rational function $h \in \overline{\mathbb{K}}(C)$ (or $\overline{\mathbb{K}}(\widetilde{C})$ likewise) is in a certain sense predetermined by the support. The following definitions are meant to make this distinction precise.

Definition 7. *Let* $h \in \overline{\mathbb{K}}(C)$ *be a rational function. We define the* twisted principal divisor $(h)_I := (h) + \widetilde{D}_I \in \mathrm{Div}(\widetilde{C})$. *With this definition the divisor* $(h)_I$ *has local equation* $h_{i,I} = u_i^{-c_i-1+\delta_I(i-1)} v_i^{-c_i-1+\delta_I(i)} h$ *in* $\pi^{-1}(C \cap U_i)$.

If h is given by an element of $\mathcal{L}_S(D_I)$, i.e. by a polynomial with support in Π_I, then the local equations $h_{i,I}$ are given by polynomials in u_i and v_i. Therefore in this case $(h)_I$ is an effective divisor on $\widetilde{C}$.

Definition 8. *Let* ν *be a valuation of* $\overline{\mathbb{K}}(C)$. *We define the* twist of the valuation ν *by* $\nu_I(h) := \nu(h_{i,I})$ *for all rational functions* $h \in \overline{\mathbb{K}}(C)$ *if* $\mathrm{Center}(\nu) \in C \cap U_i$ *and* $h_{i,I}$ *as in the previous definition.*

Taking into account the previous definition that means $\nu_I(h) = \nu(h)$ if $\mathrm{Center}(\nu) \in C \cap T$ and $\nu_I(h) = \nu(h) - \nu\left(v_i^{c_i+1-\delta_I(i)}\right)$ if $\mathrm{Center}(\nu) \in C \cap E_i$. Note that $C \cap E_i \cap E_j = \emptyset$ for $i \neq j$.

7.5.3 Adjoint divisors and a canonical divisor

Now we want to define the *adjoint order* at a point $P \in \widetilde{C}$ which is given by a valuation ν_P of $\overline{\mathbb{K}}(C)$. In sections 7.3.3 and 7.3.4 we have seen that giving ν_P is the same as giving a (primitive) injective homomorphism $\varphi_P : \overline{\mathbb{K}}(C) \rightarrow \overline{\mathbb{K}}((t))$. Assume that $\mathrm{Center}(\nu_P) \in U_i$. We define the adjoint order (see also [6, Remark 2.5] and 7.7.1 in the example section) as follows:

$$\alpha_{\nu_P} := \nu_P\left(\frac{\partial f}{\partial v_i}\right) - \mathrm{ord}_t\left(\frac{\mathrm{d}\,\varphi_P(u_i)}{\mathrm{d}\,t}\right)$$

It can be shown that this definition is indeed independent of the choice of φ_P and if Q is a smooth point of C then $\alpha_{\nu_P} = 0$. Further the adjoint order at conjugate points is the same, i.e. $\alpha_{\nu_P} = \alpha_{\nu_\sigma P}$ for $\sigma \in \mathrm{Gal}(\overline{\mathbb{K}} \mid \mathbb{K})$. So

$$\widetilde{A} := \sum_\nu \alpha_\nu P_\nu \in \mathrm{Div}(\widetilde{C})$$

where ν runs over all discrete valuations of $\overline{\mathbb{K}}(C)$ is a well defined $\mathbb{K}$-rational divisor. It is actually given by the finite sum $\sum_{\nu \in V_{\mathrm{Sing}(C)}} \alpha_\nu P_\nu$.

Definition 9. *With the above notation we define the* shifted adjoint divisor

$$\widetilde{K}_I := \widetilde{D}_I - \widetilde{A}.$$

Note that $\widetilde{K}_I$ is the difference of two $\mathbb{K}$-rational divisors and hence $\mathbb{K}$-rational as well. We know that $\deg(\widetilde{A}) = 2 \sum_{Q \in C} \delta_Q$ (see for example [8, p. 1620]). Together with lemma 6 and theorem 4 this implies:

Corollary 10. *The degree of the divisor $\widetilde{K}_I$ is* $\deg(\widetilde{K}_I) = d_I - 2$.

We have $\widetilde{C} \cong \mathbb{P}^1_{\overline{\mathbb{K}}}$ and so a divisor on $\widetilde{C}$ is *canonical* if it has degree -2. In particular $\widetilde{K}_\emptyset$ is a canonical divisor. The importance of the divisors $\widetilde{K}_I$ stems from the following theorem:

Theorem 11. *If $I \neq [1, n]$ and $d_I \geq 2$ then $\mathcal{L}_{\widetilde{C}}(\widetilde{K}_I) \neq \emptyset$ and we can compute a basis in $\mathbb{K}(C)$.*

Proof. See [1, Theorem 17].

Let us briefly recall the algorithm within the constructive proof. We start with the $\overline{\mathbb{K}}$-vector space $\mathcal{L}_S(D_I)$ (for a basis of that space see lemma 2 above) and compute the subspace $V := \{h \in \mathcal{L}_S(D_I) \mid \nu_I(h) \geq \alpha_\nu \text{ for all } \nu \in V_{\mathrm{Sing}(C)}\}$. Then $\dim_{\mathbb{K}}(V) = d_I - 1$. A priori V is a space of rational functions on the surface S, but it can be considered a space of rational functions over C or $\widetilde{C}$ as well. It turns out that via this identification $V \cong \mathcal{L}_{\widetilde{C}}(\widetilde{K}_I)$ as $\overline{\mathbb{K}}$-vector spaces. We will execute this algorithm several times on an example in section 7.7. There it will also become clear that a basis in $\mathbb{K}(C)$ can be computed.

7.5.4 Inverting the canonical divisor

We want to compute a $\overline{\mathbb{K}}$-basis for $\mathcal{L}_{\widetilde{C}}(-\widetilde{K}_\emptyset)$, the linear system associated to the *anticanonical divisor*. Therefore choose $l_1, k_1, l_2, k_2 \in \mathbb{Z}/n\mathbb{Z}$ s.t. $l_2 = k_1 + 1$, $[l_1, k_2] \neq [1, n]$ and for each pair $d_{[l_j, k_j]} \geq 2$. By what was just said we can compute elements $0 \neq g_j \in \mathcal{L}_{\widetilde{C}}(\widetilde{K}_{[l_j, k_j]})$. For every valuation ν we define the twisted orders $\beta_{j,\nu} := \nu_{[l_j, k_j]}(g_j)$. Now let $S' \subset C$ be the image of the support of the (effective) divisor $(g_1)_{[l_1, k_2]} + (g_2)_{[l_1, k_2]}$. This means the twisted orders are zero for all valuations except those of $\mathcal{V}_{S'}$. More precisely we have

$$(g_j) = -\widetilde{D}_{[l_j, k_j]} + (g_j)_{[l_j, k_j]} = -\widetilde{D}_{[l_j, k_j]} + \sum_{\nu \in \mathcal{V}_{S'}} \beta_{j,\nu} P_\nu.$$

Since $g_j \in \mathcal{L}_{\widetilde{C}}(\widetilde{K}_{[l_j, k_j]})$ we further have the inequality $\beta_{j,\nu} \geq \alpha_\nu$. We will show that $g_1 g_2$ is a good denominator for computing $\mathcal{L}_{\widetilde{C}}(-\widetilde{K}_\emptyset)$ in the sense that the according numerators are all elements of $\mathcal{L}_{\widetilde{C}}(\widetilde{K}_{[l_1, k_2]})$, a space that can again be computed by means of the above theorem (for an explicit computation see 7.7.4 in the example section).

Theorem 12. *With the choices from above define the subspace*

$$V := \{h \in \mathcal{L}_{\widetilde{C}}(\widetilde{K}_{[l_1, k_2]}) \mid \nu_{[l_1, k_2]}(h) \geq \beta_{1,\nu} + \beta_{2,\nu} - \alpha_\nu \forall \nu \in \mathcal{V}_{S'}\}.$$

Then $\mathcal{L}_{\widetilde{C}}(-\widetilde{K}_\emptyset) \to V : k \mapsto k g_1 g_2$ *is an isomorphism of* $\overline{\mathbb{K}}$*-vector spaces.*

Proof. Any rational function k can be written as $k = \frac{h}{g_1 g_2}$ for some other rational function h. Now $k \in \mathcal{L}_{\widetilde{C}}(-\widetilde{K}_\emptyset)$ if and only if

$$
\begin{aligned}
0 \leq \left(\frac{h}{g_1 g_2}\right) &- \widetilde{K}_\emptyset \\
&= (h) - (g_1) - (g_2) - \widetilde{K}_\emptyset \\
&= (h) + (\widetilde{D}_{[l_1, k_1]} + \widetilde{D}_{[l_2, k_2]} - \widetilde{D}_\emptyset) + \sum_{\nu \in \mathcal{V}_{S'}} (-\beta_{1,\nu} - \beta_{2,\nu} + \alpha_\nu) P_\nu \\
&= (h) + \underbrace{\widetilde{D}_{[l_1, k_2]} + \sum_{\nu \in \mathcal{V}_{S'}} (-\beta_{1,\nu} - \beta_{2,\nu} + \alpha_\nu) P_\nu}_{\widetilde{H}:=}
\end{aligned}
$$

if and only if $h \in \mathcal{L}_{\widetilde{C}}(\widetilde{H})$. But $-\beta_{1,\nu} - \beta_{2,\nu} + \alpha_\nu \leq -\alpha_\nu$ and therefore $\mathcal{L}_{\widetilde{C}}(\widetilde{H}) \subset \mathcal{L}_{\widetilde{C}}(\widetilde{K}_{[l_1, k_2]})$. The exact calculation above shows that $V = \mathcal{L}_{\widetilde{C}}(\widetilde{H})$.

In particular $\dim_{\overline{\mathbb{K}}}(V) = \dim_{\overline{\mathbb{K}}}(\mathcal{L}_{\widetilde{C}}(-\widetilde{K}_\emptyset)) = 3$ and if $\{b_1, b_2, b_3\} \subset V$ is a $\overline{\mathbb{K}}$-basis then $\left\{\frac{b_1}{g_1 g_2}, \frac{b_2}{g_1 g_2}, \frac{b_3}{g_1 g_2}\right\}$ is a $\overline{\mathbb{K}}$-basis of $\mathcal{L}_{\widetilde{C}}(-\widetilde{K}_\emptyset)$. In an actual computation, we will of course again start from $\mathcal{L}_S(D_{[l_1, k_2]})$ and impose directly the vanishing conditions of this theorem. When executing the algorithm in section 7.7 we will see that the output basis will already be in $\mathbb{K}(C)$.

7.6 The algorithm

We now summarize the resulting algorithm. For the input equation $f \in \mathbb{K}[x, y]$ of our algorithm we require:

Condition (*): The Newton polygon $\Pi(f)$ is non-degenerate and we can find l_1, k_1, l_2, k_2 as in section 7.5.4.

Condition (*) is fulfilled if $\Pi(f)$ is non-degenerate and all edges have length ≥ 2. If there is an edge with length $= 1$ then the algorithm of paper [1] can be used to compute a rational parametrization without field extension, because in this case it is easy to find a rational point on the orignal curve. Therefore (*) is not critical.

Some remarks to algorithm 1 are in order. The delta invariants δ_P can be computed using Puiseux expansions or Hamburger-Noether expansions (in the case of positive characteristic), see [6]. Implementations exist in Maple (characteristic zero only) and in Singular [7]. These tools also provide a way to represent the valuations ν by suitable homomorphisms into Laurent series rings.

7.7 Example

We want to parametrize the curve $C \subset \mathbb{A}^2_{\mathbb{Q}}$ given implicitly by the equation

$$f = -27y^2x^3 - 4y^3x + 13y^2x^2 + 8yx^3 - 20y^2x - 8yx^2$$
$$+ \, 4y^2 - 8yx + 4x^2 + 8y + 8x + 4.$$

Hence the field of definition is $\mathbb{Q}$. The Newton polygon $\Pi(f)$ is depicted in figure 7.4. It has 6 vertices and 4 interior points. It can be represented as the intersection of 8 half planes which are governed by the following set of data:

$v_1 = (0,0),$	$(a_1, b_1) = (0, 1),$	$c_1 = 0,$
$v_2 = (2,0),$	$(a_2, b_2) = (-1, 1),$	$c_2 = -2,$
$v_3 = (3,1),$	$(a_3, b_3) = (-1, 0),$	$c_3 = -3,$
$v_4 = (3,2),$	$(a_4, b_4) = (-1, -1),$	$c_4 = -5,$
$v_5 = (3,2),$	$(a_5, b_5) = (-1, -2),$	$c_5 = -7,$
$v_6 = (1,3),$	$(a_6, b_6) = (0, -1),$	$c_6 = -3,$
$v_7 = (1,3),$	$(a_7, b_7) = (1, -1),$	$c_7 = -2,$
$v_8 = (0,2),$	$(a_8, b_8) = (1, 0),$	$c_8 = 0$

The half planes with normals (a_4, b_4) and (a_6, b_6) have been inserted in order to fulfill equation (7.3). Hence the constructed toric surface is covered by 8 affine charts $S = \bigcup_{i \in [1,8]} U_i$. Using (7.4) we compute the equations in all the charts:

Algorithm 1 $Parametrize(f : \mathbb{K}[x,y]) : \mathbb{L}(t)^2 \cup \{\text{FAIL}\}$

Require: a polynomial $f \in \mathbb{K}[x,y]$ irreducible as an element of $\overline{\mathbb{K}}[x,y]$
        satisfying condition (*)
Ensure: a proper parametrization $(X(t), Y(t)) \in \mathbb{L}(t)^2$ s.t.
        $f(X(t), Y(t)) = 0$ or FAIL if no such parametrization exists
        (here $\mathbb{L}$ is an algebraic extension of $\mathbb{K}$ of least degree)
 1: Compute $\Pi(f)$ and determine the chart representation of the curve C embedded in the
     toric surface $S = \bigcup_{1 \leq i \leq n} U_i$ (see section 7.2.4);
 2: $\delta := 0$;
 3: **for** $P \in \text{Sing}(C)$ **do**
 4:    Compute the delta invariant δ_P;
 5:    $\delta := \delta + \delta_P$;
 6: **end for**
 7: **if** $\#(\Pi(f)^\circ) - \delta \neq 0$ **then**
 8:    **return** FAIL; {The genus is not zero, see theorem 4.}
 9: **end if**
10: Find l_1, k_1, l_2, k_2 as in section 7.5.4;
11: Set $S_1 := \mathcal{L}_S(D_{[l_1,k_1]})$, $S_2 := \mathcal{L}_S(D_{[l_2,k_2]})$ and $S_3 := \mathcal{L}_S(D_{[l_1,k_2]})$ (see lemma 2);
12: Compute α_ν for every valuation $\nu \in \mathcal{V}_{\text{Sing}(C)}$ (see section 7.5.3);
13: **for** $\nu \in \mathcal{V}_{\text{Sing}(C)}$ **do**
14:    Set $S_1 := \{g \in S_1 \mid \nu_{[l_1,k_1]}(g) \geq \alpha_\nu\}$;
15:    Set $S_2 := \{g \in S_2 \mid \nu_{[l_2,k_2]}(g) \geq \alpha_\nu\}$;
     {These steps compute $\mathcal{L}_{\widetilde{C}}(\widetilde{K}_{[l_1,k_1]})$ and $\mathcal{L}_{\widetilde{C}}(\widetilde{K}_{[l_2,k_2]})$ using theorem 11.}
16: **end for**
17: Choose elements $0 \neq g_1 \in S_1$ and $0 \neq g_2 \in S_2$, compute the intersection locus $I \subset C$
     of the effective divisor $(g_1) + D_{[l_1,k_1]} + (g_2) + D_{[l_2,k_2]}$ with the curve C and the values
     $\beta_{j,\nu}$ for $\nu \in \mathcal{V}_I$ (see section 7.5.3);
18: **for** $\nu \in \mathcal{V}_I$ **do**
19:    Set $S_3 := \{g \in S_3 \mid \nu_{[l_1,k_2]}(g) \geq \beta_{1,\nu} + \beta_{2,\nu} + \alpha_\nu\}$;
20: **end for**
21: Choose a basis $\{b_1, b_2, b_3\} \subset S_3 \cap \mathbb{K}[x,y]$.
22: Compute the defining equation of the image $C' \subset \mathbb{A}^2_{\mathbb{K}}$ of the birational map $\psi_1 : (x,y) \mapsto$
     $(\frac{b_1}{b_3}, \frac{b_2}{b_3})$ and compute its rational inverse ψ_1^{-1}.
23: Find a parametrization ψ_2 of the conic C' using a minimal degree field extension.
24: **return** $\psi_1^{-1} \circ \psi_2$;

$$f_1 = -27v_1^2u_1^3 - 4v_1^3u_1 + 13v_1^2u_1^2 + 8v_1u_1^3 - 20v_1^2u_1 - 8v_1u_1^2$$
$$+ \, 4v_1^2 - 8v_1u_1 + 4u_1^2 + 8v_1 + 8u_1 + 4$$
$$f_2 = -4v_2^4u_2^3 + 4v_2^4u_2^2 - 20v_2^3u_2^2 + 8v_2^3u_2 + 13v_2^2u_2^2 - 8v_2^2u_2$$
$$- \, 27v_2u_2^2 + 4v_2^2 - 8v_2u_2 + 8v_2 + 8u_2 + 4$$
$$f_3 = \ldots$$

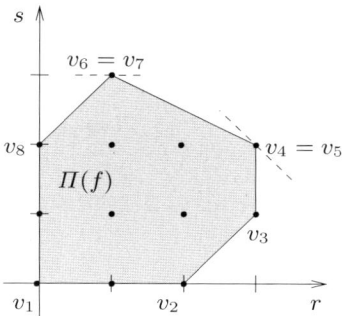

Fig. 7.4. Newton polygon of example

7.7.1 Analysis of the singularities

The first chart U_1 contains two singular points, Q_1 with coordinates $(u_1, v_1) = (-1, 0)$ and Q_2 with coordinates $(u_1, v_1) = (0, -1)$. The point Q_1 is also contained in U_2 and Q_2 is contained in U_8. These are already all singular points, so all subsequent computations can be done in the first chart. For future reference we also give the partial derivative by v_1:

$$\frac{\mathrm{d}\,f_1}{\mathrm{d}\,v_1} = -54v_1^2 u_1^3 - 12v_1^3 u_1 + 26v_1^2 u_1^2 + 8v_1 u_1^3 - 40v_1^2 u_1 - 8v_1 u_1^2$$
$$+ \; 8v_1^2 - 8v_1 u_1 + 8v_1$$

According to the singularities we compute Puiseux expansions in $u_1 + 1$ for Q_1 and in u_1 for Q_2:

$$\sigma_1(u_1 + 1) = -\tfrac{1}{4}(u_1 + 1) + \gamma_1(u_1 + 1)^2 + (\tfrac{435}{608}\gamma_1 - \tfrac{51}{2432})(u_1 + 1)^3 \ldots$$
$$\sigma_2(u_1) = -1 - \tfrac{5}{2}u_1 + \gamma_2 u_1^2 + (\tfrac{21}{4}\gamma_2 + \tfrac{195}{16})u_1^3 \ldots$$

Here γ_1 and γ_2 are elements of $\overline{\mathbb{Q}}$ s.t. $1024\gamma_1^2 + 516\gamma_1 + 63 = 0$ and $16\gamma_2^2 + 24\gamma_2 - 45 = 0$. From these expansions we get the following two monomorphisms from the function field into a field of Laurent series:

$$\varphi_1 : Q(\overline{\mathbb{Q}}[u_1, v_1]/f_1) \to \overline{\mathbb{Q}}((t)) :$$
$$\begin{cases} u_1 + 1 \mapsto t \\ v_1 \mapsto -\tfrac{1}{4}t + \gamma_1 t^2 + (\tfrac{435}{608}\gamma_1 - \tfrac{51}{2432})t^3 \ldots \end{cases}$$
$$\varphi_2 : Q(\overline{\mathbb{Q}}[u_1, v_1]/f_1) \to \overline{\mathbb{Q}}((t)) :$$
$$\begin{cases} u_1 \mapsto t \\ v_1 + 1 \mapsto -\tfrac{5}{2}t + \gamma_2 t^2 + (\tfrac{21}{4}\gamma_2 + \tfrac{195}{16})t^3 \ldots \end{cases}$$

These homomorphisms induce valuations $\nu_i := \mathrm{ord}_t \circ \varphi_i$. Using these valuations we are able to speak about a resolution $\pi : \tilde{C} \to C$ without constructing it explicitly.

For example ν_1 and its conjugate with respect to the Galois extension $\mathbb{Q}(\gamma_1) \mid \mathbb{Q}$ correspond to two points in the preimage $\pi^{-1}(Q_1)$. We compute the adjoint orders for these valuations (and their conjugates):

$$\begin{aligned}
\alpha_{\nu_1} &= \mathrm{ord}_t\left(\varphi_1\left(\frac{\mathrm{d}\,f_1}{\mathrm{d}\,v_1}\right)\right) - \mathrm{ord}_t\left(\frac{\mathrm{d}\,\varphi_1(u_1)}{\mathrm{d}\,t}\right) \\
&= \mathrm{ord}_t\left(\left(\tfrac{129}{4} + 128\gamma_1\right)t^2 + \left(-\tfrac{3225}{76} - \tfrac{3200}{19}\gamma_1\right)t^3 + \ldots\right) - \mathrm{ord}_t(1) \\
&= 2 - 0 = 2 \\
\alpha_{\nu_2} &= \mathrm{ord}_t\left(\varphi_2\left(\frac{\mathrm{d}\,f_1}{\mathrm{d}\,v_1}\right)\right) - \mathrm{ord}_t\left(\frac{\mathrm{d}\,\varphi_2(u_1)}{\mathrm{d}\,t}\right) \\
&= \mathrm{ord}_t\left((6 + 8\gamma_2)t^2 + \left(\tfrac{39}{2} + 26\gamma_2\right)t^3 + \ldots\right) - \mathrm{ord}_t(1) \\
&= 2 - 0 = 2
\end{aligned}$$

7.7.2 Checking rationality

The number of interior points of $\Pi(f)$ is equal to 4. Again from [8, p. 1620], we see that $\delta_{Q_1} = \delta_{Q_2} = 2$. Hence $\mathrm{genus}(C) = 4 - 2 - 2 = 0$ and C is parametrizable.

7.7.3 Computing two adjoints

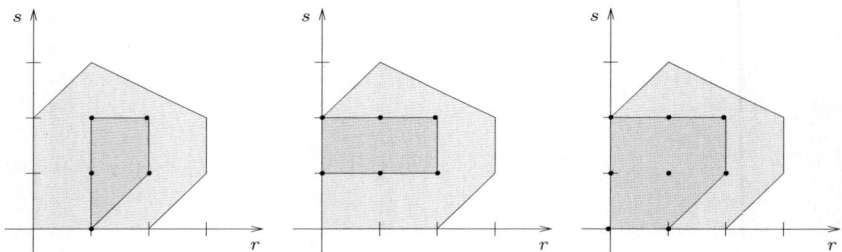

Fig. 7.5. Bases for the linear systems $\mathcal{L}_S(D_{[1,1]})$, $\mathcal{L}_S(D_{[7,8]})$ and $\mathcal{L}_S(D_{[7,1]})$

First we have to compute a denominator $g_1 g_2$. We choose $l_1 = k_1 = 1$ and $l_2 = 7$, $k_2 = 8$. We make an indetermined Ansatz for $g_1 \in \mathcal{L}_{\tilde{C}}(\tilde{K}_{[1,1]})$ and show its local equation $g_{1,1,[1,1]}$ (compare figure 7.5):

$$g_1 := c_0 x + c_1 xy + c_2 x^2 y + c_3 xy^2 + c_4 x^2 y^2$$
$$g_{1,1,[1,1]} = c_0 + c_1 v_1 + c_2 u_1 v_1 + c_3 v_1^2 + c_4 u_1 v_1^2$$

Requiring $\nu_1(g_{1,1,[1,1]}) \geq \alpha_{\nu_1}$ and $\nu_2(g_{1,1,[1,1]}) \geq \alpha_{\nu_2}$ yields the following constraints:

$$0 = c_0, \qquad\qquad\qquad 0 = \tfrac{1}{4}c_2 - \tfrac{1}{4}c_1,$$
$$0 = c_0 - c_1 + c_3, \qquad\qquad 0 = -c_2 - \tfrac{5}{2}c_1 + 5c_3 + c_4$$

Note that this step would in general yield linear constraints over a field extension, e.g. over $\mathbb{Q}(\gamma_1)$ or $\mathbb{Q}(\gamma_2)$. They can always be written as a bigger number of constraints over the ground field, here $\mathbb{Q}$. This is the algorithmic side of the statement that the divisor $\widetilde{K}_{[1,1]}$ is $\mathbb{Q}$-rational. Solving the system with respect to c_3 and setting $c_3 = 1$, we get $g_1 = xy + x^2y + xy^2 - \tfrac{3}{2}x^2y^2$. We do the same again for $g_2 \in \mathcal{L}_{\widetilde{C}}(\widetilde{K}_{[7,8]})$:

$$g_2 := c_0 y + c_1 xy + c_2 x^2 y + c_3 y^2 + c_4 xy^2 + c_5 x^2 y^2$$
$$g_{2,1,[7,8]} = c_0 + c_1 u_1 + c_2 u_1^2 + c_3 v_1 + c_4 u_1 v_1 + c_5 u_1^2 v_1$$

Requiring $\nu_1(g_{2,1,[7,8]}) \geq \alpha_{\nu_1}$ and $\nu_2(g_{2,1,[7,8]}) \geq \alpha_{\nu_2}$ yields the following constraints:

$$0 = c_0 - c_1 + c_2, \qquad\qquad 0 = \tfrac{1}{4}c_4 - 2c_2 - \tfrac{1}{4}c_5 - \tfrac{1}{4}c_3 + c_1,$$
$$0 = c_0 - c_3, \qquad\qquad\qquad 0 = -c_4 - \tfrac{5}{2}c_3 + c_1$$

Solving the system with respect to c_3 and c_4 and setting $c_3 = 2$ and $c_4 = -1$ (which is a matter of choice), we get $g_2 = 2y + 4xy + 2x^2y + 2y^2 - xy^2 - 3x^2y^2$.

The system for computing g_1 had got an 1-dimensional solution. This implies that the divisor $(g_1)_{[1,1]}$ is equal to $\widetilde{A}$. Hence its support is exactly the preimage of the singular locus. On the other hand we were left with essentially one degree of freedom when computing g_2 and therefore $(g_2)_{[7,8]} > \widetilde{A}$. Indeed we find that the support of g_2 has an additional point corresponding to $Q_3 \in U_1$, namely $(u_1, v_1) = (-1, -16)$. Therefore we have to consider one more valuation ν_3 centered at Q_3 (which one could get again from a Puiseux series solution at Q_3). All this is reflected in the twisted orders of g_1 and g_2:

$$\beta_{1,\nu_1} = \nu_{1,[1,1]}(g_1) = 2, \quad \beta_{1,\nu_2} = \nu_{2,[1,1]}(g_1) = 2, \quad \beta_{1,\nu_3} = \nu_{3,[1,1]}(g_1) = 0,$$
$$\beta_{2,\nu_1} = \nu_{1,[7,8]}(g_2) = 2, \quad \beta_{2,\nu_1} = \nu_{2,[7,8]}(g_2) = 2, \quad \beta_{2,\nu_3} = \nu_{3,[7,8]}(g_2) = 1$$

Since $Q_3 \in C$ is a smooth point, the adjoint order is $\alpha_{\nu_3} = 0$.

7.7.4 Linear system of an anticanonical divisor

Now we make an Ansatz for an element $h \in \mathcal{L}_{\widetilde{C}}(\widetilde{K}_{[7,1]})$:

$$h := c_1 + c_2 x + c_3 y + c_4 xy + c_5 x^2 y + c_6 y^2 + c_7 xy^2 + c_8 x^2 y^2$$
$$h_{1,[7,1]} := c_1 + c_2 u_1 + c_3 v_1 + c_4 u_1 v_1 + c_5 u_1^2 v_1 + c_6 v_1^2 + c_7 u_1 v_1^2 + c_8 u_1^2 v_1^2$$

Requiring $\nu_1(h_{1,[7,1]}) \geq \beta_{1,\nu_1} + \beta_{2,\nu_1} - \alpha_{\nu_1} = 2$, $\nu_2(h_{1,[7,1]}) \geq \beta_{1,\nu_2} + \beta_{2,\nu_2} - \alpha_{\nu_2} = 2$ and $\nu_3(h_{1,[7,1]}) \geq \beta_{1,\nu_3} + \beta_{2,\nu_3} - \alpha_{\nu_3} = 1$ (i.e. $h_{1,[7,1]}$ also has to vanish on Q_3) yields the following constraints:

$$0 = c_1 - c_2$$
$$0 = -\tfrac{1}{4}c_5 + \tfrac{1}{4}c_4 + c_2 - \tfrac{1}{4}c_3$$
$$0 = c_1 - c_3 + c_6$$
$$0 = c_7 - c_4 + c_2 + 5c_6 - \tfrac{5}{2}c_3$$
$$0 = c_1 - c_2 - 16c_3 + 16c_4 - 16c_5 + 256c_6 - 256c_7 + 256c_8$$

Solving the system with respect to c_6, c_7 and c_8, we compute a basis of the vector space V of theorem 12:

$$b_1 = 4 + 4x + 5y - \tfrac{7}{2}yx + \tfrac{15}{2}yx^2 + y^2$$
$$b_2 = -4 - 4x - 4y + 7yx - 5yx^2 + y^2 x$$
$$b_3 = 4 + 4x + 4y - 6yx + 6yx^2 + y^2 x^2$$

In other words $\left\{\frac{b_1}{g_1 g_2}, \frac{b_2}{g_1 g_2}, \frac{b_3}{g_1 g_2}\right\}$ is a basis of $\mathcal{L}_{\widetilde{C}}(-\widetilde{K}_\emptyset)$, the linear system of our anticanonical divisor.

7.7.5 Birational equivalence to a conic

The rational functions $\frac{b_i}{g_1 g_2}$ are the coordinates of a map from C to the projective plane $\mathbb{P}^2_{\mathbb{Q}}$. We get the same map if we multiply all coordinates by their common denominator $g_1 g_2$:

$$C \dashrightarrow \mathbb{P}^2_{\mathbb{Q}} : \begin{pmatrix} x \\ y \end{pmatrix} \mapsto \begin{pmatrix} 4 + 4x + 5y - \tfrac{7}{2}yx + \tfrac{15}{2}yx^2 + y^2 \\ -4 - 4x - 4y + 7yx - 5yx^2 + y^2 x \\ 4 + 4x + 4y - 6yx + 6yx^2 + y^2 x^2 \end{pmatrix}$$

Dividing by the last coordinate, we get a map to $\mathbb{A}^2_{\mathbb{Q}}$:

$$\psi_1 : C \dashrightarrow \mathbb{A}^2_{\mathbb{Q}} : \begin{pmatrix} x \\ y \end{pmatrix} \mapsto \begin{pmatrix} \frac{8+8x+10y-7yx+15yx^2+2y^2}{2(4+4x+4y-6yx+6yx^2+y^2x^2)} \\ \frac{-4-4x-4y+7yx-5yx^2+y^2x}{4+4x+4y-6yx+6yx^2+y^2x^2} \end{pmatrix}$$

The image of this map is a conic $C' \subset \mathbb{A}^2_{\mathbb{Q}}$. To avoid confusion, we use the coordinates x' and y' in the image domain. We can compute the implicit equation $f' = 12x'y' + y'^2 + 9x'$ of C' by eliminating the variables x and y using Gröbner bases techniques. Then ψ_1 is a birational morphism with inverse

$$\psi_1^{-1} : C' \dashrightarrow C : \begin{pmatrix} x' \\ y' \end{pmatrix} \mapsto \begin{pmatrix} \frac{(4y'+3)(4y'^2+36y'+27)}{y'(2y'+3)(34y'+27)} \\ \frac{-(2y'+3)(14y'^2+45y'+27)}{4y'^2(4y'+3)} \end{pmatrix}.$$

In other words ψ_1^{-1} is a parametrization of C by the conic C'.

7.7.6 Parametrization of the conic

The conic C' can be parametrized easily (e.g. by stereographic projection) if one point on it is known. In our case the origin is contained in C' and we compute the following parametrization over $\mathbb{Q}$ using a pencil of lines through the origin:

$$\psi_2 : \mathbb{A}^1_{\mathbb{Q}} \dashrightarrow C' : t \mapsto \begin{pmatrix} \frac{-9}{t^2+12t} \\ \frac{-9t}{t^2+12t} \end{pmatrix}$$

Composing both maps finally yields a parametrization of the input curve:

$$\psi_1^{-1} \circ \psi_2 : \mathbb{A}^1_{\mathbb{Q}} \dashrightarrow C : t \mapsto \begin{pmatrix} \frac{-(12+12t+t^2)t}{3(2+3t)(6+t)} \\ \frac{-(6+9t+t^2)(6+t)}{12t} \end{pmatrix}$$

Acknowledgments

The authors were supported by the FWF (Austrian Science Fund) in the frame of the research projects SFB 1303.

References

1. Tobias Beck and Josef Schicho. Parametrization of Algebraic Curves Defined by Sparse Equations. *AAECC*, 2006. Accepted for publication. Preprint available as RICAM Report 2005–08 at http://www.ricam.oeaw.ac.at/publications/reports.
2. Wieb Bosma, John Cannon, and Catherine Playoust. The Magma algebra system. I. The user language. *J. Symbolic Comput.*, 24(3-4):235–265, 1997. Computational algebra and number theory (London, 1993).
3. David Cox. What is a toric variety? In *Topics in Algebraic Geometry and Geometric Modeling*, volume 334 of *Contemporary Mathematics*, pages 203–223. American Mathematical Society, Providence, Rhode Island, 2003. Workshop on Algebraic Geometry and Geometric Modeling (Vilnius, 2002).
4. David A. Cox. Toric varieties and toric resolutions. In *Resolution of singularities (Obergurgl, 1997)*, volume 181 of *Progr. Math.*, pages 259–284. Birkhäuser, Basel, 2000.
5. William Fulton. *Introduction to toric varieties*, volume 131 of *Annals of Mathematics Studies*. Princeton University Press, Princeton, NJ, 1993. The William H. Roever Lectures in Geometry.
6. G.-M. Greuel, C. Lossen, and M. Schulze. Three algorithms in algebraic geometry, coding theory and singularity theory. In *Applications of algebraic geometry to coding theory, physics and computation (Eilat, 2001)*, volume 36 of *NATO Sci. Ser. II Math. Phys. Chem.*, pages 161–194. Kluwer Acad. Publ., Dordrecht, 2001.
7. Gert-Martin Greuel and Gerhard Pfister. *A Singular introduction to commutative algebra*. Springer-Verlag, Berlin, 2002. With contributions by Olaf Bachmann, Christoph Lossen and Hans Schönemann, With 1 CD-ROM (Windows, Macintosh, and UNIX).
8. Gaétan Haché and Dominique Le Brigand. Effective construction of algebraic geometry codes. *IEEE Trans. Inform. Theory*, 41(6, part 1):1615–1628, 1995. Special issue on algebraic geometry codes.

9. Robin Hartshorne. *Algebraic geometry*. Springer-Verlag, New York, 1977. Graduate Texts in Mathematics, No. 52.
10. F. Hess. Computing Riemann-Roch spaces in algebraic function fields and related topics. *J. Symbolic Comput.*, 33(4):425–445, 2002.
11. Erik Hillgarter and Franz Winkler. Points on algebraic curves and the parametrization problem. In *Wang, Dongming (ed.), Automated deduction in geometry. International workshop, Toulouse, France, September 27–29, 1996. Proceedings. Berlin: Springer. Lect. Notes Comput. Sci. 1360, 189-207* . 1998.
12. Nathan Jacobson. *Finite-dimensional division algebras over fields*. Springer-Verlag, Berlin, 1996.
13. Josef Schicho. On the choice of pencils in the parametrization of curves. *J. Symbolic Comput.*, 14(6):557–576, 1992.
14. J. Rafael Sendra and Franz Winkler. Symbolic parametrization of curves. *J. Symbolic Comput.*, 12(6):607–631, 1991.
15. J. Rafael Sendra and Franz Winkler. Parametrization of algebraic curves over optimal field extensions. *J. Symbolic Comput.*, 23(2-3):191–207, 1997. Parametric algebraic curves and applications (Albuquerque, NM, 1995).
16. Igor R. Shafarevich. *Basic algebraic geometry. 1*. Springer-Verlag, Berlin, second edition, 1994. Varieties in projective space, Translated from the 1988 Russian edition and with notes by Miles Reid.
17. Denis Simon. Solving quadratic equations using reduced unimodular quadratic forms. *Math. Comp.*, 74(251):1531–1543 (electronic), 2005.
18. Mark van Hoeij. Rational parametrizations of algebraic curves using a canonical divisor. *J. Symbolic Comput.*, 23(2-3):209–227, 1997. Parametric algebraic curves and applications (Albuquerque, NM, 1995).
19. Oscar Zariski and Pierre Samuel. *Commutative algebra. Vol. II*. Springer-Verlag, New York, 1975. Reprint of the 1960 edition, Graduate Texts in Mathematics, No. 29.

8

Ridges and Umbilics of Polynomial Parametric Surfaces

Frédéric Cazals[1], Jean-Charles Faugère[2], Marc Pouget[3], and Fabrice Rouillier[2]

[1] INRIA Sophia-Antipolis, Geometrica project,
    2004 route des Lucioles, BP 93,
    F-06902 Sophia-Antipolis, FRANCE.
    Frederic.Cazals@sophia.inria.fr
[2] INRIA Rocquencourt and
    Universit Pierre et Marie Curie-Paris6, UMR 7606, LIP6, Salsa project,
    Domaine de Voluceau, BP 105,
    F-78153 Le Chesnay Cedex, FRANCE.
    Jean-Charles.Faugere@inria.fr
    Fabrice.Rouillier@inria.fr
[3] LORIA, INRIA Nancy - Grand Est,
    Villers-lès-Nancy, F-54602 FRANCE.
    Marc.Pouget@loria.fr

Summary. Given a smooth surface, a blue (red) ridge is a curve such that at each of its point, the maximum (minimum) principal curvature has an extremum along its curvature line. As curves of *extremal* curvature, ridges are relevant in a number of applications including surface segmentation, analysis, registration, matching. In spite of these interests, given a smooth surface, no algorithm reporting a certified approximation of its ridges was known so far, even for restricted classes of generic surfaces.

This paper partly fills this gap by developing the first algorithm for polynomial parametric surfaces — a class of surfaces ubiquitous in CAGD. The algorithm consists of two stages. First, a polynomial bivariate implicit characterization of ridges $P = 0$ is computed using an implicitization theorem for ridges of a parametric surface. Second, the singular structure of $P = 0$ is exploited, and the approximation problem is reduced to solving zero dimensional systems using Rational Univariate Representations. An experimental section illustrates the efficiency of the algorithm on Bézier patches.

8.1 Introduction

8.1.1 Ridges

Originating with the parabolic lines drawn by Felix Klein on the Apollo of Belvedere [10], curves on surfaces have been a natural way to apprehend the aesthetics of shapes [12]. Aside these artistic concerns, applications such as surface segmentation, analysis, registration or matching [11, 16] are concerned with the curves of *extremal curvature* of a surface, which are its so-called *ridges*. (We note in passing that interestingly, (selected) ridges are also central in the analysis of Delaunay based surface meshing algorithms [1].)

A comprehensive literature on ridges exists – see [11, 17, 18], and we just introduce the basic notions so as to discuss our contributions. Consider a smooth embedded surface whose principal curvatures are denoted k_1 and k_2 with $k_1 \geq k_2$. Away from umbilical points — where $k_1 = k_2$, principal directions of curvature are well defined, and we denote them d_1 and d_2. In local coordinates, we denote $\langle , \rangle$ the inner product induced by the ambient Euclidean space, and the gradients of the principal curvatures are denoted dk_1 and dk_2. Ridges can be defined as follows — see Fig. 8.1 for an illustration :

Definition 1. *A non umbilical point is called*

- *a blue ridge point if the* extremality coefficient $b_0 = \langle dk_1, d_1 \rangle$ *vanishes, i.e.* $b_0 = 0$.
- *a red ridge point if the* extremality coefficient $b_3 = \langle dk_2, d_2 \rangle$ *vanishes, i.e.* $b_3 = 0$.

As the principal curvatures are not differentiable at umbilics, note that the extremality coefficients are not defined at such points. Notice also the sign of the extremality coefficients is not defined, as each principal direction can be oriented by two opposite unit vectors. Apart from umbilics, special points on ridges are *purple* points – they actually correspond to intersections between red and a blue ridges. The calculation of ridges poses difficulties of three kinds.

Topological difficulties.

Ridges of a smooth surface feature self-intersections at umbilics — more precisely at so-called 3-ridges umbilics — and purple points. From a topological viewpoint, reporting a certified approximation of ridges therefore requires reporting these singular points.

Numerical difficulties.

As ridges are characterized by derivatives of principal curvatures, reporting them requires evaluating third order differential quantities. Estimating such derivatives depends upon the particular type of surface processed — implicitly defined, parameterized, discretized by a mesh, but is numerically a demanding task.

Orientation difficulties.

As observed above, the signs of the b_0 and b_3 depend upon the particular orientations of the principal directions picked. But as a global coherent non vanishing orientation of the principal directions cannot be found in the neighborhoods of umbilics, tracking the zero crossings of b_0 and b_3 faces a major difficulty. For the particular case of surfaces represented by meshes, the so-called *Acute rule* can be used [4], but computing meshes compliant with the requirements imposed by the acute rule is an open problem. For surfaces represented implicitly or parametrically, one can resort to the Gaussian extremality $E_g = b_0 b_3$, which eradicates the sign problems, but prevents from reporting the red and blue ridges separately.

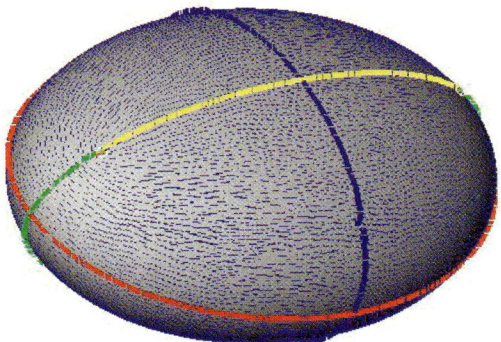

Fig. 8.1. Umbilics, ridges, and principal blue foliation on the ellipsoid for normals pointing outward

8.1.2 Previous work

Given the previous difficulties, no algorithm reporting ridges in a certified fashion had been developed until this work. Most contributions deal with sampled surfaces known through a mesh, and a complete review of these contributions can be found in [4]. In the following, we focus on contributions related to parametric surfaces.

Reporting umbilics. Umbilics of a surface are always traversed by ridges so that reporting ridges faithfully requires reporting umbilics. To do so, Morris [13] minimizes the function $k_1 - k_2$, which vanishes exactly at umbilics. Meakawa et al. [15] define a polynomial system whose roots are the umbilics. This system is solved with the *rounded interval arithmetic projected polyhedron method*. This algorithm uses specific properties of the Bernstein basis of polynomials and interval arithmetic. The domain is recursively subdivided and a set of boxes containing the umbilics is output, but neither existence nor uniqueness of an umbilic in a box is guaranteed.

Reporting ridges. The only method dedicated to parametric surfaces we are aware of is that of Morris [13, 14]. The parametric domain is triangulated and zero crossings are sought on edges. Local orientation of the principal directions are needed but only provided with a heuristic. This enables to detect crossings assuming (i) there is at most one such crossing on an edge (ii) the orientation of the principal directions is correct. As this simple algorithm fails near umbilics, these points are located first and crossings are found on a circle around the umbilic.

Equation of the ridge curve. Ridges can be characterized either as extrema of principal curvatures along their curvature lines as in definition 1, or by analyzing the contact between the surface and spheres [11]. For parametric surfaces, this later approach allows a global characterization of ridges [18, Chapter 11] as a 1 dimensional smooth sub-manifold in a 7 dimensional space. But this characterization is not amenable to algorithmic developments.

    Shifting from this seven-dimensional space to the parametric space, the theory of algebraic invariants has been used to derive the equation of the ridge curve as the zero set of an invariant function [8]. The ensuing strategy consists of defining invariants as functions of the fundamental forms and their derivatives. The equation of ridges is given in this setting. If one further specializes this equation for a surface given by a parameterization, the result matches, up to a constant factor, our implicit encoding $P = 0$ [3]. The point of view of our approach is to work from the beginning on a parametrized surface. The definition of ridges involves principal curvatures and principal directions of curvature which are independent of the given parametrization, but we explicit all these invariants wrt the parametrization and its derivatives. Hence, for polynomial parametric surfaces, we end with a polynomial with integer coefficients whose variables are the partial derivatives of the parametrization up to the third order. This polynomial is the same for any other parametrization.

Reporting the topology of an algebraic curve. In the case of a polynomial parametric surface, we recast the problem of approximating ridges into the field of algebraic geometry. We recall that the standard tool to compute a graph encoding the topology of a 2-D or 3-D curve is the Cylindrical Algebraic Decomposition (CAD) [7, 9].

8.1.3 Contributions and paper overview

Let $\Phi(u, v)$ be a smooth parameterized surface over a domain $\mathcal{D} \subset \mathbb{R}^2$. We wish to report a certified approximation of its ridges, which subsumes a solution for all the difficulties enumerated in section 8.1.1.

    The first step in providing a certified approximation of the ridges of Φ consists of computing an implicit equation $P = 0$ encoding these ridges. The derivation of this equation is presented in the companion paper [3], which also contains a detailed discussion of our implicit encoding of ridges wrt previous work.

    The equation $P = 0$ being taken for granted, the contribution developed in this paper is to exploit as far as possible the geometry of $\mathcal{P}$ encoded in $P = 0$, so as to develop the first algorithm able to compute the ridges topology of a polynomial parametric surface. Our algorithm avoids the main difficulties of CAD methods: (i) singular and critical points are sequentially computed directly in $2D$; (ii) no generic assumption is required, i.e. several critical or singular points may have the same horizontal projection; (iii) no computation with algebraic numbers is involved. Because algorithms based on the Cylindrical Algebraic Decomposition are not effective for our high degree curves such as $P = 0$, our algorithm is to the best of our knowledge the only one able to certify properties of the curve $P = 0$.

    The paper is organized as follows. The implicit equations for ridges and its singularities are recalled in section 8.2. The algorithm to compute the topology of the ridge curve is described in section 8.3. Section 8.4 provides illustrations on two Bézier surfaces.

8.1.4 Notations

Ridges and umbilics.

At any non umbilical point of the surface, the maximal (minimal) principal curvature is denoted k_1 (k_2), and its associated direction d_1 (d_2). Anything related to the maximal (minimal) curvature is qualified blue (red), for example we shall speak of the blue curvature for k_1 or the red direction for d_2. Since we shall make precise statements about ridges, it should be recalled that, according to definition 1, umbilics are not ridge points.

Differential calculus.

For a bivariate function $f(u,v)$, the partial derivatives are denoted with indices, for example $f_{uuv} = \frac{\partial^3 f}{\partial^2 u \partial v}$. The gradient of f is denoted f_1 or $df = (f_u, f_v)$. The quadratic form induced by the second derivatives is denoted $f_2(u,v) = f_{uu}u^2 + 2f_{uv}uv + f_{vv}v^2$. The discriminant of this form is denoted $\delta(f_2) = f_{uv}^2 - f_{uu}f_{vv}$. The cubic form induced by the third derivatives in denoted $f_3(u,v) = f_{uuu}u^3 + 3f_{uuv}u^2v + 3f_{uvv}uv^2 + f_{vvv}v^3$. The discriminant of this form is denoted $\delta(f_3) = 4(f_{uuu}f_{uvv} - f_{uuv}^2)(f_{uuv}f_{vvv} - f_{uvv}^2) - (f_{uuu}f_{vvv} - f_{uuv}f_{uvv})^2$.

Let f be a real bivariate polynomial and $\mathcal{F}$ the real algebraic curve defined by f. A point $(u,v) \in \mathbb{C}^2$ is called

- a singular point of $\mathcal{F}$ if $f(u,v) = 0$, $f_u(u,v) = 0$ and $f_v(u,v) = 0$;
- a critical point of $\mathcal{F}$ if $f(u,v) = 0$, $f_u(u,v) = 0$ and $f_v(u,v) \neq 0$ (such a point has an horizontal tangent, we call it critical because if one fixes the v coordinate, then the restricted function is critical wrt the u coordinate, this notion will be useful in section 8.3);
- a regular point of $\mathcal{F}$ if $f(u,v) = 0$ and it is neither singular nor critical.

If the domain $\mathcal{D}$ of study is a subset of $\mathbb{R}^2$, one calls fiber a cross section of this domain at a given ordinate or abscissa.

Misc.

The inner product of two vectors x, y is denoted $\langle x, y \rangle$.

8.2 Relevant equations for ridges and its singularities

This section briefly recalls the equations defining the ridge curve and its singularities, see [3]. Let Φ be the parameterization of class C^k for $k \geq 4$. Denote I and II the matrices of the first and second fundamental form of the surface in the basis (Φ_u, Φ_v) of the tangent space. In order for normals and curvatures to be well defined, we assume the surface is regular i.e. $\det(I) \neq 0$.

The principal directions d_i and principal curvatures $k_1 \geq k_2$ are the eigenvectors and eigenvalues of the matrix $W = I^{-1}II$. The following equation defines coefficients A, B, C and D as polynomials wrt the derivative of the parameterization Φ up to the second order

$$\begin{pmatrix} A & B \\ C & D \end{pmatrix} = W(\det I)^{3/2}. \tag{8.1}$$

As a general rule, in the following calculations, we will be interested in deriving quantities which are polynomials wrt the derivatives of the parameterization. These calculations are based on quantities (principal curvatures and directions) which are independent of a given parameterization, hence the derived formula are valid for any parameterization.

Umbilics are characterized by the equation $p_2 = 0$, with $p_2 = (k_1 - k_2)^2 (\det I)^3$. We then define two vector fields v_1 and w_1 orienting the principal direction field d_1

$$v_1 = (-2B, A - D - \sqrt{p_2})$$
$$w_1 = (A - D + \sqrt{p_2}, 2C).$$

Derivatives of the principal direction k_1 wrt these two vector fields define a, a', b, b' by the equations:

$$a\sqrt{p_2} + b = \sqrt{p_2}(\det I)^{5/2}\langle dk_1, v_1 \rangle \; ; \quad a'\sqrt{p_2} + b' = \sqrt{p_2}(\det I)^{5/2}\langle dk_1, w_1 \rangle. \tag{8.2}$$

The following definition is a technical tool to state the next theorem in a simple way. The function $Sign_{ridge}$ introduced here will be used to classify ridge colors. Essentially, this function describes all the possible sign configurations for ab and $a'b'$ at a ridge point.

Definition 2. *The function $Sign_{ridge}$ takes the values*

$$-1 \;\; if \begin{cases} ab < 0 \\ a'b' \leq 0 \end{cases} \;\; or \;\; \begin{cases} ab \leq 0 \\ a'b' < 0 \end{cases},$$

$$+1 \;\; if \begin{cases} ab > 0 \\ a'b' \geq 0 \end{cases} \;\; or \;\; \begin{cases} ab \geq 0 \\ a'b' > 0 \end{cases},$$

$$0 \;\; if \, ab = a'b' = 0.$$

Theorem 3. *The set of blue ridges union the set of red ridges union the set of umbilics has equation $P = 0$ where $P = (a^2 p_2 - b^2)/B$ is a polynomial wrt $A, B, C, D, \det I$ as well as their first derivatives and hence is a polynomial wrt the derivatives of the parameterization up to the third order. For a point of this set $\mathcal{P}$, one has:*

- *If $p_2 = 0$, the point is an umbilic.*
- *If $p_2 \neq 0$ then:*
  - *if $Sign_{ridge} = -1$ then the point is a blue ridge point,*
  - *if $Sign_{ridge} = +1$ then the point is a red ridge point,*
  - *if $Sign_{ridge} = 0$ then the point is a purple point.*

In addition, the classification of an umbilic as 1-ridge or 3-ridges from P_3 goes as follows:

- If P_3 is elliptic, that is the discriminant of P_3 is positive ($\delta(P_3) > 0$), then the umbilic is a 3-ridge umbilic and the 3 tangent lines to the ridges at the umbilic are distinct.
- If P_3 is hyperbolic ($\delta(P_3) < 0$) then the umbilic is a 1-ridge umbilic.

8.2.1 Polynomial surfaces

A fundamental class of surface used in Computer Aided Geometric Design consists of polynomial surfaces like Bézier and splines. We first observe that if Φ is a polynomial, all its derivatives are also polynomials. Thus in the polynomial case the equation of ridges, which is a polynomial wrt to these derivatives, is algebraic. Hence the set of all ridges and umbilics is globally described by an algebraic curve. Notice that the parameterization can be general, in which case $\Phi(u, v) = (x(u, v), y(u, v), z(u, v))$, or can be a height function $\Phi(u, v) = (u, v, z(u, v))$.

As a corollary of Thm. 3, one can give upper bounds for the total degree of the polynomial P wrt that of the parameterization. Distinguishing the cases where Φ is a general parameterization or a height function (that is $\Phi(u, v) = (u, v, h(u, v))$) with $h(u, v)$ and denoting d the total degree of Φ, P has total degree $33d - 40$ or $15d - 22$ for a height function.

In the more general case where the parameterization is given by rational fractions of polynomials, P is a rational function of the surface parameters too. The denominator of P codes the points where the surface is not defined and away from these points, the numerator codes the ridges and umbilics.

8.3 Certified topological approximation

In this section, we circumvent the difficulties of the Cylindrical Algebraic Decomposition (CAD) and develop a certified algorithm to compute the topology of $\mathcal{P}$. Consider a parameterized surface $\Phi(u, v)$, the parameterization being *polynomial* with rational coefficients. Let $\mathcal{P}$ be the curve encoding the ridges of $\Phi(u, v)$. We aim at studying $\mathcal{P}$ on the compact box domain $\mathcal{D} = [a, b] \times [c, d]$.

Given a real algebraic curve, the standard way to approximate it consists of resorting to the CAD. Running the CAD requires computing singular points and critical points of the curve — points with a horizontal tangent. Theoretically, these points are defined by zero-dimensional systems. Practically, because of the high degree of the polynomials involved, the calculations may not go through. Replacing the bottlenecks of the CAD by a resolution method adapted to the singular structure of $\mathcal{P}$, we develop an algorithm producing a graph $\mathcal{G}$ embedded in the domain $\mathcal{D}$, which is isotopic to the curve $\mathcal{P}$ of ridges in $\mathcal{D}$. Key points are that:

1. no generic assumption is required, i.e. several critical or singular points may have the same horizontal projection;
2. no computation with algebraic numbers is involved.

8.3.1 Algebraic tools

Two algebraic methods are ubiquitously called by our algorithm: univariate root isolation and rational univariate representation. We briefly present these tools and give references for the interested reader.

Univariate root isolation. This tool enables to isolate roots of univariate polynomials whose coefficients are rational numbers, by means of intervals with rational bounds. The method uses the Descartes rule and is fully explained in [20].

Rational univariate representation [19]. The Rational Univariate Representation is, with the end-user point of view, the simplest way for representing symbolically the roots of a zero-dimensional system without loosing information (multiplicities or real roots) since one can get all the information on the roots of the system by solving univariate polynomials.

Given a zero-dimensional system

$$I = < p_1, \ldots, p_s >$$

where the $p_i \in \mathbb{Q}[X_1, \ldots, X_n]$, a Rational Univariate Representation of $V(I)$, has the following shape:

$$f_t(T) = 0, X_1 = \frac{g_{t,X_1}(T)}{g_{t,1}(T)}, \ldots, X_n = \frac{g_{t,X_n}(T)}{g_{t,1}(T)},$$

where $f_t, g_{t,1}, g_{t,X_1}, \ldots, g_{t,X_n} \in \mathbb{Q}[T]$ (T is a new variable). It is uniquely defined w.r.t. a given polynomial t which separates $V(I)$ (injective on $V(I)$), the polynomial f_t being necessarily the characteristic polynomial of m_t in $\mathbb{Q}[X_1, \ldots, X_n]/I$. The RUR defines a one-to-one map between the roots of I and those of f_t preserving the multiplicities and the real roots:

$$
\begin{array}{ccc}
V(I)(\cap \mathbb{R}) & \approx & V(f_t)(\cap \mathbb{R}) \\
\alpha = (\alpha_1, \ldots, \alpha_n) & \rightarrow & t(\alpha) \\
\left(\frac{g_{t,X_1}(t(\alpha))}{g_{t,1}(t(\alpha))}, \ldots, \frac{g_{t,X_n}(t(\alpha))}{g_{t,1}(t(\alpha))} \right) & \leftarrow & t(\alpha)
\end{array}
$$

The RUR also enables efficient evaluation of the sign of polynomials at the roots of a system.

8.3.2 Assumptions on the ridge curve and study points

According to the structure of the singularities of the ridge curve recalled in section 8.2, the only assumption made is that the surface admits generic ridges in the sense that real singularities of $\mathcal{P}$ satisfy the following conditions:

- Real singularities of $\mathcal{P}$ are of multiplicity at most 3.
- Real singularities of multiplicity 2 are called purple points. They satisfy the system $S_p = \{a = b = a' = b' = 0, \ \delta(P_2) > 0, \ p_2 \neq 0\}$. In addition, this implies that two real branches of $\mathcal{P}$ are passing through a purple point.

- Real singularities of multiplicity 3 are called umbilics and they satisfy the system $S_u = \{p_2 = 0\} = \{p_2 = 0, P = 0, P_u = 0, P_v = 0\}$. In addition, if $\delta(P_3)$ denote the discriminant of the cubic of the third derivatives of P at an umbilic, one has:
  - if $\delta(P_3) > 0$, then the umbilic is called a 3-ridge umbilic and three real branches of $\mathcal{P}$ are passing through the umbilic with three distinct tangents;
  - if $\delta(P_3) < 0$, then the umbilic is called a 1-ridge umbilic and one real branch of $\mathcal{P}$ is passing through the umbilic.

As we shall see in section 8.3.5, these conditions are checked during the processing of the algorithm.

Given this structure of singular points, the algorithm successively isolate umbilics, purple points and critical points. As a system defining one set of these points also includes the points of the previous system, we use a localization method to simplify the calculations. The points reported at each stage are characterized as roots of a zero-dimensional system — a system with a finite number of complex solutions, together with the number of half-branches of the curve connected to each point. In addition, points on the border of the domain of study need a special care. This setting leads to the definition of *study points*:

Definition 4. *Study points are points in $\mathcal{D}$ which are*

- *real singularities of $\mathcal{P}$, that is $S_s = S_u \cup S_p$, with $S_u = S_{1R} \cup S_{3R}$ and*
  - $S_{1R} = \{p_2 = P = P_u = P_v = 0, \delta(P_3) < 0\}$
  - $S_{3R} = \{p_2 = P = P_u = P_v = 0, \delta(P_3) > 0\}$
  - $S_p = \{a = b = a' = b' = 0, \delta(P_2) > 0, p_2 \neq 0\}$
    $= \{a = b = a' = b' = 0, \delta(P_2) > 0\} \setminus S_u$
- *real critical points of $\mathcal{P}$ in the v-direction (i.e. points with a horizontal tangent which are not singularities of $\mathcal{P}$) defined by the system*
  $S_c = \{P = P_u = 0, P_v \neq 0\}$;
- *intersections of $\mathcal{P}$ with the left and right sides of the box $\mathcal{D}$ satisfying the system*
  $S_b = \{P(a, v) = 0, v \in [c, d]\} \cup \{P(b, v) = 0, v \in [c, d]\}$. *Such a point may also be critical or singular.*

8.3.3 Output specification

Definition 5. *Let $\mathcal{G}$ be a graph whose vertices are points of $\mathcal{D}$ and edges are non-intersecting straight line-segments between vertices. Let the topology on $\mathcal{G}$ be induced by that of $\mathcal{D}$. We say that $\mathcal{G}$ is a topological approximation of the ridge curve $\mathcal{P}$ on the domain $\mathcal{D}$ if $\mathcal{G}$ is ambient isotopic to $\mathcal{P} \cap \mathcal{D}$ in $\mathcal{D}$.*
More formally, there exists a function $F : \mathcal{D} \times [0, 1] \longrightarrow \mathcal{D}$ such that:

- *F is continuous;*
- *$\forall t \in [0, 1]$, $F_t = F(., t)$ is an homeomorphism of $\mathcal{D}$ onto itself;*
- *$F_0 = Id_{\mathcal{D}}$ and $F_1(\mathcal{P} \cap \mathcal{D}) = \mathcal{G}$.*

Note that homeomorphic approximation is weaker and our algorithm actually gives isotopy. In addition, our construction allows to identify singularities of $\mathcal{P}$ to a subset of vertices of $\mathcal{G}$ while controlling the error on the geometric positions. We can also color edges of $\mathcal{G}$ with the color of the ridge curve it is isotopic to. Once this topological sketch is given, one can easily compute a more accurate geometrical picture.

8.3.4 Method outline

Taking the square free part of P, we can assume P is square free. We can also assume $\mathcal{P}$ has no part which is a horizontal segment — parallel to the u-axis. Otherwise this means that a whole horizontal line is a component of P. In other words, the content of P wrt u is a polynomial in v and we can study this factor separately and divide P by this factor. Eventually, to get the whole topology of the curve, one has to merge the components.

Our algorithms consists of the following five stages:

1. **Isolating study points.** Study point are isolated in $2D$ with rational univariate representations (RUR). Study points within a common fiber are identified.
2. **Regularization of the study boxes.** We know the number of branches of the curve going through each study point. The boxes of study points are reduced so as to be able to define the number of branches coming from the bottom and from the top.
3. **Computing regular points in study fibers.** In each fiber of a study point, the u-coordinates of intersection points with $\mathcal{P}$ other than study points are computed.
4. **Adding intermediate rational fibers.** Add rational fibers between study points fibers and isolate the u-coordinates of intersection points with $\mathcal{P}$.
5. **Performing connections.** This information is enough to perform the connections. Consider the cylinder between two consecutive fibers, the number of branches connected from above the lower fiber is the same than the number of branches connected from below the higher fiber. Hence there is only one way to perform connections with non-intersecting straight segments.

8.3.5 Step 1. Isolating study points

The method to identify these study points is to compute a RUR of the system defining them. More precisely, we sequentially solve the following systems:

1. The system S_u from which the sets S_{1R} and S_{3R} are distinguished by evaluating the sign of $\delta(P_3)$.
2. The system S_p for purple points.
3. The system S_c for critical points.
4. The system S_b for border points, that is intersections of $\mathcal{P}$ with the left and right sides of the box $\mathcal{D}$. Solving this system together with one of the previous identifies border points which are also singular or critical.

Selecting only points belonging to $\mathcal{D}$ reduces to adding inequalities to the systems and is well managed by the RUR. According to [19], solving such systems is equivalent to solving zero-dimensional systems without inequalities when the number of inequations remains small compared to the number of variables. The RUR of the study points provides a way to compute a box around each study point q_i which is a product of two intervals $[u_i^1; u_i^2] \times [v_i^1; v_i^2]$. The intervals can be as small as desired.

Until now, we only have separate information on the different systems. In order to identify study points having the same v-coordinate, we need to cross this information. First we compute isolation intervals for all the v-coordinates of all the study points together, denote I this list of intervals. If two study points with the same v-coordinate are solutions of two different systems, the gcd of polynomials enable to identify them:

- Initialize the list I with all the isolation intervals of all the v-coordinates of the different systems.
- Let A and B be the square free polynomials defining the v-coordinates of two different systems, and I_A, I_B the lists of isolation intervals of their roots. Let $C = gcd(A, B)$ and I_C the list of isolation intervals of its roots. One can refine the elements of I_C until they intersect only one element of I_A and one element of I_B. Then replace these two intervals in I by the single interval which is the intersection of the three intervals. Do the same for every pair of systems.
- I then contains intervals defining different real numbers in one-to-one correspondence with the v-coordinates of the study points. It remains to refine these intervals until they are all disjoint.

Second, we compare the intervals of I and those of the 2d boxes of the study points. Let two study points q_i and q_j be represented by $[u_i^1; u_i^2] \times [v_i^1; v_i^2]$ and $[u_j^1; u_j^2] \times [v_j^1; v_j^2]$ with $[v_i^1; v_i^2] \cap [v_j^1; v_j^2] \neq \varnothing$. One cannot, a priori, decide if these two points have the same v-coordinate or if a refinement of the boxes will end with disjoint v-intervals. On the other hand, with the list I, such a decision is straightforward. The boxes of the study points are refined until each $[v_i^1; v_i^2]$ intersects only one interval $[w_i^1; w_i^2]$ of the list I. Then two study points intersecting the same interval $[w_i^1; w_i^2]$ are in the same fiber.

Finally, one can refine the u-coordinates of the study points with the same v coordinate until they are represented with disjoint intervals since, thanks to localizations, all the computed points are distinct.

Checking genericity conditions of section 8.3.2.

First, real singularities shall be the union of purple and umbilical points, this reduces to compare the systems for singular points and for purple and umbilical points. Second, showing that $\delta(P_3) \neq 0$ for umbilics and $\delta(P_2) > 0$ for purple points reduces to sign evaluation of polynomials at the roots of a system (see section 8.3.1).

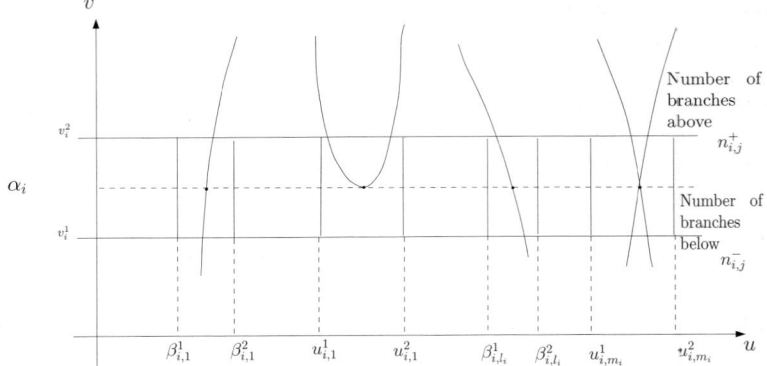

Fig. 8.2. Notations for a fiber involving several critical/singular points: $u_{i,j}^{1(2)}$ are used for study points, $\beta_{i,j}^{1(2)}$ for simple points.

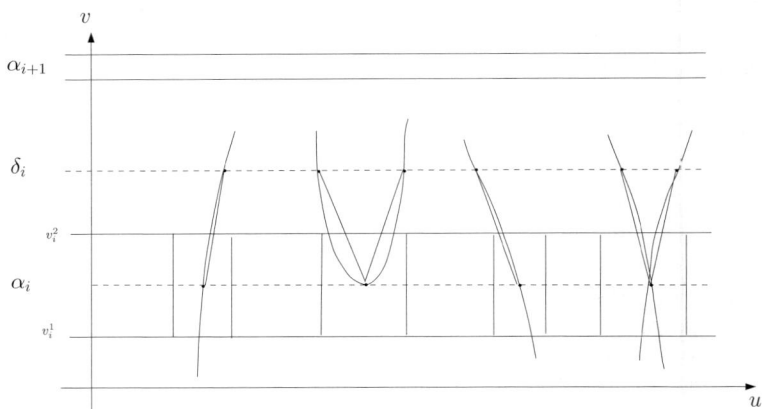

Fig. 8.3. Performing connections between the study point fiber α_i and the intermediate fiber δ_i

8.3.6 Step 2. Regularization of the study boxes

At this stage, we have computed isolating boxes of all study points $\{q_{i,j}, i = 1 \ldots s, j = 1 \ldots m_i\}$: the v-coordinates $\alpha_1, \ldots, \alpha_s$ are isolated by intervals $[v_i^1; v_i^2]$, $i = 1 \ldots s$ and the u-coordinates of the m_i study points in each fiber α_i are isolated by intervals $[u_{i,j}^1; u_{i,j}^2], j = 1 \ldots m_i$.

We know the number of branches of the curve passing through each study point : it is 6 for a 3-ridge umbilic, 4 for a purple and 2 for others. We want to compute the number branches coming from the bottom and from the top. We first reduce the box until the number of intersections between the curve and the border of the box matches the known number of branches connected to the study point. Then the intersections are obviously in one-to-one correspondence with the branches. Second, as in [21] for example, we reduce the height of the box again if necessary so that intersections only occur on the top or the bottom of the box.

Counting the number of intersections reduces to solve 4 univariate polynomials with rational coefficients. Reducing a box means refining its representation with the RUR.

8.3.7 Step 3. Computing regular points in study fibers

We now compute the regular points in each fiber $P(u, \alpha_i) = 0$. Computing the regular points of each fiber is now equivalent to computing the roots of the polynomials $P(u, \alpha_i)$ outside the intervals representing the u-coordinates of the study points (which contain all the multiple roots of $P(u, \alpha_i)$).

Denote $[u_{i,j}^1; u_{i,j}^2], j = 1..m_i$ the intervals representing the u-coordinates of the study points on the fiber of α_i and $[v_i^1, v_i^2]$ an interval containing (strictly) α_i and no other $\alpha_j, j \neq i$. Substituting v by any rational value $q \in [v_i^1, v_i^2]$ in $P(u, v)$ gives a univariate polynomial with rational coefficients $P(u, q)$. We then isolate the (simple) roots of this polynomial $P(u, q)$ on the domain $[a, b] \setminus \cup_{j=1}^{m_i} [u_{i,j}^1; u_{i,j}^2]$: the algorithm returns intervals $[\beta_{i,j}^1; \beta_{i,j}^2], j = 1 \dots l_i$ representing these roots.

To summarize the information up to this point : we have, along each fiber, a collection of points $s_{i,j}$, $i = 1 \dots s$, $j = 1, \dots, m_i + l_i$, which are either study points or regular points of $\mathcal{P}$. Each such point is isolated in a box i.e. a product of intervals and comes with two integers $(n_{i,j}^+, n_{i,j}^-)$ denoting the number of branches in $\mathcal{D}$ connected from above and from below.

8.3.8 Step 4. Adding intermediate rational fibers

Consider now an intermediate fiber, i.e. a fiber associated with $v = \delta_i\ i = 1 \dots s - 1$, with δ_i a rational number in-between the intervals of isolation of two consecutive values α_i and α_{i+1}. If the fibers $v = c$ or $v = d$ are not fibers of study points, then they are added as fibers δ_0 or δ_s.

Getting the structure of such fibers amounts to solving a univariate polynomial with rational coefficients, which is done using the algorithm described in section 8.3.1. Thus, each such fiber also comes with a collection of points, isolated in boxes, for which one knows that $n_{i,j}^+ = n_{i,j}^- = 1$.

8.3.9 Step 5. Performing connections

We thus obtain a full and certified description of the fibers: all the intersection points with $\mathcal{P}$ and their number of branches connected. We know, by construction, that the branches of $\mathcal{P}$ between fibers have empty intersection. The number of branches connected from above a fiber is the same than the number of branches connected from below the next fiber. Hence there is only one way to perform connections with non-intersecting straight segments. More precisely, vertices of the graph are the centers of isolation boxes, and edges are line-segments joining them.

Notice that using the intermediate fibers $v = \delta_i$ is compulsory if one wishes to get a graph $\mathcal{G}$ isotopic to $\mathcal{P}$. If not, whenever two branches have common starting points and endpoints, the embedding of the graph $\mathcal{G}$ obtained is not valid since two arcs are identified.

The algorithm is illustrated on Fig. 8.3. In addition

- If a singular point box have width δ, then the distance between the singular point and the vertex representing it is less than δ.
- One can compute the sign of the function $Sign_{ridge}$ (definition 2) for each regular point of each intermediate fiber. This defines the color of the ridge branch it belongs to. Then one can assign to each edge of the graph the color of its end point which is on an intermediate fiber.

8.4 Illustration

We provide the topology of ridges for two Bézier surfaces defined over the domain $\mathcal{D} = [0,1] \times [0,1]$.

The first surface has control points

$$\begin{pmatrix} [0,0,0] & [1/4,0,0] & [2/4,0,0] & [3/4,0,0] & [4/4,0,0] \\ [0,1/4,0] & [1/4,1/4,1] & [2/4,1/4,-1] & [3/4,1/4,-1] & [4/4,1/4,0] \\ [0,2/4,0] & [1/4,2/4,-1] & [2/4,2/4,1] & [3/4,2/4,1] & [4/4,2/4,0] \\ [0,3/4,0] & [1/4,3/4,1] & [2/4,3/4,-1] & [3/4,3/4,1] & [4/4,3/4,0] \\ [0,4/4,0] & [1/4,4/4,0] & [2/4,4/4,0] & [3/4,4/4,0] & [4/4,4/4,0] \end{pmatrix}$$

Alternatively, this surface can be expressed as the graph of the total degree 8 polynomial $h(u,v)$ for $(u,v) \in [0,1]^2$:

$$h(u,v) = 116u^4v^4 - 200u^4v^3 + 108u^4v^2 - 24u^4v - 312u^3v^4 + 592u^3v^3 - 360u^3v^2$$
$$+80u^3v + 252u^2v^4 - 504u^2v^3 + 324u^2v^2 - 72u^2v - 56uv^4 + 112uv^3 - 72uv^2 + 16uv.$$

The computation of the implicit curve has been performed using Maple 9.5 and requires less than one minute (see [3]). It is a bivariate polynomial $P(u,v)$ of total degree 84, of degree 43 in u, degree 43 in v with 1907 terms and coefficients with up to 53 digits. Figure 8.4 displays the topological approximation graph of the ridge curve in the parametric domain $\mathcal{D}$ computed with the algorithm of section 8.3. There are 19 critical points, 17 purple points and 8 umbilics, 3 of which are 3-ridge and 5 are 1-ridge.

We have computed the subsets S_u, S_p and S_c by using the software FGB and RS (http://fgbrs.lip6.fr). The RUR can be computed as shown in [19] or alternatively, Gröbner basis can be computed first using [5] or [6]. We tested both methods and the computation time for the biggest system S_c does not exceed 10 minutes with a Pentium M 1.6 Ghz. The following table gives the main characteristics of these systems:

System	# of roots $\in \mathbb{C}$	# of roots $\in \mathbb{R}$	# of real roots $\in \mathcal{D}$
S_u	160	16	8
S_p	749	47	17
S_c	1432	44	19

In order to have more insight of the geometric meaning of the ridge curve, the surface and its ridges are displayed on Fig. 8.5. This plot is computed without topological certification with the `rs_tci_points` function (from RS software, see also [2]) from the polynomial P and then lifted on the surface.

Fig. 8.4. Bi-quartic Bèzier example : isotopic approximation of the ridge curve with study points circled.

The second surface is a bi-quadratic Bézier surface

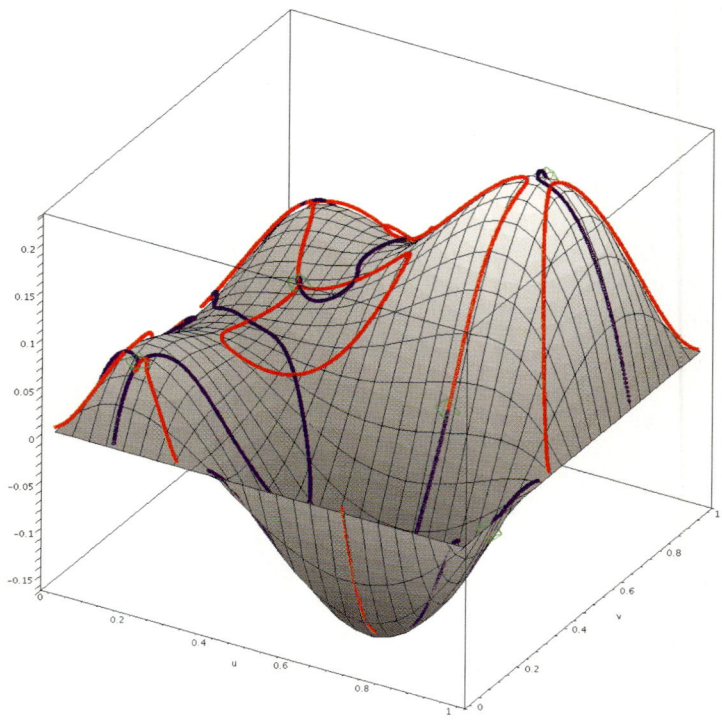

Fig. 8.5. Plot of the bi-quartic Bèzier surface with ridges

$$\Phi(u, v) = [2/3\,v + 2/3\,uv - 1/3\,u^2 v + 1/3\,v^2 - 2/3\,v^2 u + 1/3\,u^2 v^2,$$
$$1/2\,u + 1/2\,u^2 + uv - u^2 v - 1/2\,v^2 u + 1/2\,u^2 v^2,$$
$$1 + 3\,v^2 - u - 4\,v + 5\,uv + u^2 v - 7/2\,v^2 u - 5/2\,u^2 v^2]$$

The ridge curve has total degree 56 and partial degrees 33 with 1078 terms and coefficients with up to 15 digits. The computation of the biggest system of study points S_c takes 4.5 minutes. On this example, study point boxes have to be refined up to a size of less than 2^{-255} to compute the topology. The following table gives the main characteristics of the study point systems:

System	# of roots $\in \mathbb{C}$	# of real roots $\in \mathcal{D}$
S_u	70	1
S_p	293	6
S_c	695	5

Figure 8.6 displays the topology of the ridges. In addition to study points, the regular points of all fibers are displayed as small black dots.

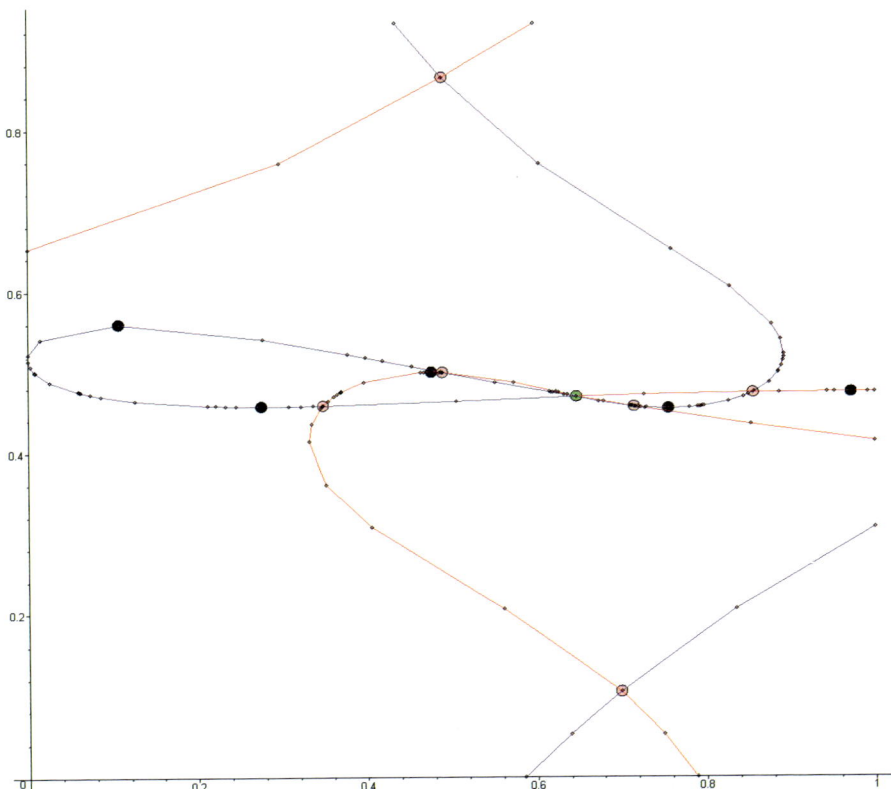

Fig. 8.6. Bi-quadratic Bèzier example : isotopic approximation of the ridge curve with study points.

8.5 Conclusion

For parametric algebraic surfaces, we developed an algorithm to report a topologically certified approximation of the ridges. This algorithm is computationally demanding in terms of algebra. It is in a sense complementary to the heuristic one developed in a companion paper [4], which is working directly on a triangulation of the surface, and provide a fast way to report non certified results.

The method developed for the computation of the topology of the ridges can be generalized for other algebraic curves. It gives an alternative to usual algorithms based on the CAD provided one knows the geometry of curve branches at singularities.

Acknowledgments

F. Cazals and M. Pouget acknowledge the support of the AIM@Shape and ACS European projects. Jean-Pierre Merlet is acknowledged for fruitful discussions.

References

1. D. Attali, J.-D. Boissonnat, and A. Lieutier. Complexity of the delaunay triangulation of points on surfaces the smooth case. In *ACM SoCG*, San Diego, 2003.
2. F. Cazals, J.-C. Faugère, M. Pouget, and F. Rouillier. Topologically certified approximation of umbilics and ridges on polynomial parametric surface. Technical Report 5674, INRIA, 2005.
3. F. Cazals, J.-C. Faugère, M. Pouget, and F. Rouillier. The implicit structure of ridges of a smooth parametric surface. *Computer Aided Geometrc Design*, 23(7):582–598, 2006.
4. F. Cazals and M. Pouget. Topology driven algorithms for ridge extraction on meshes. Technical Report RR-5526, INRIA, 2005.
5. J.-C. Faugère. A new efficient algorithm for computing gröbner bases (f_4). *Journal of Pure and Applied Algebra*, 139(1-3):61–88, June 1999.
6. J.-C. Faugère. A new efficient algorithm for computing gröbner bases without reduction to zero f_5. In *International Symposium on Symbolic and Algebraic Computation Symposium - ISSAC 2002, Villeneuve d'Ascq, France*, Jul 2002.
7. G. Gatellier, A. Labrouzy, B. Mourrain, and J.-P. Tècourt. Computing the topology of 3-dimensional algebraic curves. In *Computational Methods for Algebraic Spline Surfaces*, pages 27–44. Springer-Verlag, 2004.
8. J. Gravesen. Third order invariants of surfaces. In T. Dokken and B. Juttler, editors, *Computational methods for algebraic spline surfaces*. Springer, 2005.
9. L. Gonzalez-Vega and I. Necula. Efficient topology determination of implicitly defined algebraic plane curves. *Computer Aided Geometric Design*, 19(9), 2002.
10. D. Hilbert and S. Cohn-Vossen. *Geometry and the Imagination*. Chelsea, 1952.
11. P. W. Hallinan, G. Gordon, A.L. Yuille, P. Giblin, and D. Mumford. *Two-and Three-Dimensional Patterns of the Face*. A.K.Peters, 1999.
12. J.J. Koenderink. *Solid Shape*. MIT, 1990.
13. R. Morris. *Symmetry of Curves and the Geometry of Surfaces: two Explorations with the aid of Computer Graphics*. Phd Thesis, 1990.
14. R. Morris. The sub-parabolic lines of a surface. In Glen Mullineux, editor, *Mathematics of Surfaces VI, IMA new series 58*, pages 79–102. Clarendon Press, Oxford, 1996.
15. T. Maekawa, F. Wolter, and N. Patrikalakis. Umbilics and lines of curvature for shape interrogation. *Computer Aided Geometric Design*, 13:133–161, 1996.
16. X. Pennec, N. Ayache, and J.-P. Thirion. Landmark-based registration using features identified through differential geometry. In I. Bankman, editor, *Handbook of Medical Imaging*. Academic Press, 2000.

17. I. Porteous. The normal singularities of a submanifold. *J. Diff. Geom.*, 5, 1971.
18. I. Porteous. *Geometric Differentiation (2nd Edition)*. Cambridge University Press, 2001.
19. F. Rouillier. Solving zero-dimensional systems through the rational univariate representation. *Journal of Applicable Algebra in Engineering, Communication and Computing*, 9(5):433–461, 1999.
20. F. Rouillier and P. Zimmermann. Efficient isolation of polynomial real roots. *Journal of Computational and Applied Mathematics*, 162(1):33–50, 2003.
21. R. Seidel and N. Wolpert. On the exact computation of the topology of real algebraic curves. In *SCG '05: Proceedings of the twenty-first annual symposium on Computational geometry*, pages 107–115, New York, NY, USA, 2005. ACM Press.
22. J.-P. Thirion. The extremal mesh and the understanding of 3d surfaces. *International Journal of Computer Vision*, 19(2):115–128, August 1996.

9

Intersecting Biquadratic Bézier Surface Patches

Stéphane Chau[1], Margot Oberneder[2], André Galligo[1], and Bert Jüttler[2]

[1] Laboratoire J.A. Dieudonné, Université de Nice - Sophia-Antipolis, France
`{chaus,galligo}@math.unice.fr`
[2] Institute of Applied Geometry, Johannes Kepler University, Austria
`{margot.oberneder,bert.juettler}@jku.at`

Summary. We present three symbolic–numeric techniques for computing the intersection and self–intersection curve(s) of two Bézier surface patches of bidegree (2,2). In particular, we discuss algorithms, implementation, illustrative examples and provide a comparison of the methods.

9.1 Introduction

The intersection of two surfaces is one of the fundamental operations in Computer Aided Design (CAD) and solid modeling. Closely related to it, the elimination of self–intersections (which may arise. e.g., from offsetting) is needed to maintain the correctness of a CAD model. Tensor–product Bézier surface patches, which are parametric surfaces defined by vector–valued polynomials $x : [0,1]^2 \to \mathbb{R}^3$ of certain bidegree (m, n), are extensively used to model surfaces in CAD and solid modeling. However, even for relatively small bidegrees $m, n \leq 3$, the intersection and self–intersection loci of such patches can be fairly complicated. Consequently, standard algorithms for surface–surface intersections [24, 28] generally do not take the properties of special classes of such tensor–product surfaces into account.

In the case of two general surfaces, a *brute–force approach* to compute the intersection curve(s) consists in (step 1) approximating the surface by triangular meshes and (step 2) intersecting the planar facets of these meshes. Clearly, in order to achieve high accuracy, a very fine approximation with a mesh may be needed. Alternatively, one may consider to choose another, more complicated representation, where the basic elements are capable of capturing more of the geometric features. For instance, one may choose quadratic triangular patches or biquadratic tensor–product patches[3]. Clearly, this approach would need robust intersection algorithms for the more complicated basic elements.

---

[3] In the same spirit, Reference [32] proposes to use triangular patches for efficient visualization.

In this paper we address the computation of the intersection curve of two surface patches of bidegree (2,2), i.e., biquadratic tensor–product patches. Our aim is to
compute the intersection by using – as far as possible – *symbolic* techniques, in order
to avoid problems with numerical robustness.

We chose the tensor–product representation, since it is more common in CAD environment. Approximations of general tensor–product surfaces by biquadratic ones
can easily be generated by combining degree reduction techniques with subdivision.
The techniques presented in this paper can immediately be extended to the case of triangular patches. Indeed, tringular patches can be seen as degenerate tensor–product
patches, where one edge collapses into a single point.

The remainder of the paper is organized as follows. After some preliminaries,
Sections 9.3 to 9.5 present three different techniques for computing the intersection
curves, which are based on *resultants*, on *approximate implicitization* (which was one
of the main research topics in the GAIA II project), and on *intersections of parameter
lines*, respectively. Section 9.6 discusses the computation of self–intersections. We
apply the three techniques to three representative examples and report the results in
Section 9.7. Finally, we conclude this paper.

9.2 Intersection and self–intersection curves

We consider the intersection curves of two biquadratic Bézier surfaces $\mathbf{x}(u, v)$ and
$\mathbf{y}(r, s)$, both with parameter domains $[0, 1]^2$. They are assumed to be given by their
parametric representations with rational coefficients (control points). More precisely,
these representations have the form

$$\mathbf{x}(u, v) = \sum_{i=0}^{2} \sum_{j=0}^{2} \mathbf{c}_{i,j} B_i(u) B_j(v) \tag{9.1}$$

with certain rational control points $\mathbf{c}_{i,j} \in \mathbb{Q}^3$ and the quadratic Bernstein polynomials $B_j(t) = \binom{2}{i} t^i (1 - t)^{2-i}$ (and similarly for the second patch $\mathbf{y}(r, s)$).

The intersection curve is defined by the system of three non–linear equations

$$\mathbf{x}(u, v) = \mathbf{y}(r, s) \tag{9.2}$$

which defines the intersection as a curve (in the generic case) in $[0, 1]^4$. Similarly,
self intersections of one of the patches are characterized by

$$\mathbf{x}(u, v) = \mathbf{x}(\bar{u}, \bar{v}). \tag{9.3}$$

In this case, the set of solutions contains the 2–plane $u = u^*$, $v = v^*$ as a trivial
component.

While these equations could be solved by using numerical methods, we plan to
explore how far it is possible to compute the intersections by using *symbolic* computations, in order to avoid rounding errors and robustness problems.

The "generic" algorithm for computing the (self–) intersection curve(s), consists
of three steps:

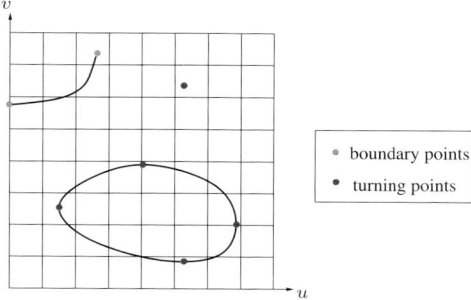

Fig. 9.1. Intersection curves in one of the parameter domains.

1. Find at least one point on each component of the intersection,
2. trace the segments of the intersection curve, and
3. collect and convert the segments into a format that is suitable for further processing (depending on the application).

We will focus on the first step, since the second step is a standard numerical problem, and step 3 depends on the specific background of the problem. Several parts of
the intersection curve may exist. Some possible types are shown in Fig. 9.1 in the
parameter domain of a Bézier surface $\mathbf{x}(u, v)$. Points with horizontal or vertical tangent are called *turning points*, and intersections with the boundaries of the patches
generate *boundary points*. Note that also isolated points (where both surfaces touch
each other) may exist.

9.3 A resultant–based approach

In this section, we will use the resultant to compute the intersection locus between
$\mathbf{x}(u, v)$ and $\mathbf{y}(r, s)$. We consider the algebraic variety

$$\mathcal{C} = \{(u, v, r, s) \mid \mathbf{x}(u, v) = \mathbf{y}(r, s)\} \tag{9.4}$$

and we will suppose that $\mathcal{C} \cap [0, 1]^4$ is a curve.

9.3.1 Resultant basics

Let f_1, f_2 and f_3 be three polynomials in two variables with given monomial supports and N the number of terms of these 3 supports. For each $i \in \{1, 2, 3\}$ we denote
by coeffs(f_i) the sequence of the coefficients of f_i. The resultant of f_1, f_2 and f_3 is,
by definition, an irreducible polynomial R in N variables with the property, that

$$R(\text{coeffs}(f_1), \text{coeffs}(f_2), \text{coeffs}(f_3)) = 0 \tag{9.5}$$

if and only if these 3 polynomials have a common root in a specified domain $\mathcal{D}$. For
a more precise description of resultants, see e.g. [2, 8, 9].

In our application to surface–surface–intersections, the resultant can be used as a projection operator. Indeed, if f_1, f_2 and f_3 are the three components of $\mathbf{x}(u, v) - \mathbf{y}(r, s)$ which are considered as polynomials in the two variables r and s, then the resultant of f_1, f_2 and f_3 is a polynomial $R(u, v)$ and it gives an implicit plane curve which corresponds to the projection of $\mathcal{C}$ in the (u, v) parameters. More precisely, if f_1, f_2 and f_3 are generic, then the two sets

$$\left\{ (u, v) \in [0, 1]^2 \mid R(u, v) = 0 \right\} \tag{9.6}$$

and

$$\left\{ (u, v) \in [0, 1]^2 \mid \exists (r, s) \in \mathcal{D} : \mathbf{x}(u, v) = \mathbf{y}(r, s) \right\} \tag{9.7}$$

are identical. Several families of multivariate resultants have been studied and some implementations are available, see [5, 22].

9.3.2 Application to the intersection problem

A strategy to describe the intersection between $\mathbf{x}(u, v)$ and $\mathbf{y}(r, s)$ consists in projecting $\mathcal{C}$ on a plane (by using the resultant). Many authors propose to project $\mathcal{C}$ on the (u, v) (or (r, s)) plane and then the resulted plane curve is traced (see [16] and [20] for the tracing method) and is lifted to the 3D space by the corresponding parameterization. Note that this method can give some unwanted components (the so called "phantom components") which are not in $\mathbf{x}([0, 1]^2) \cap \mathbf{y}([0, 1]^2)$. So, another step is needed to cut off the extraneous branches. This last part can be done with a solver for multivariate polynomial systems (see [25]) or an inversion of parameterization (see [3]).

As an alternative to these existing approaches, we propose to project the set $\mathcal{C}$ onto the (u, r) space. Note that, in the equations $\mathbf{x}(u, v) = \mathbf{y}(r, s)$, the two variables v and s are separated, so they can be eliminated via a simple resultant computation. It turns out that such a resultant can be computed via the determinant of a Bezoutian matrix (see [15]). First, consider the $(3, 3)$ determinant:

$$b = \det \left(\mathbf{x}(u, v) - \mathbf{y}(r, s), \frac{\mathbf{x}(u, v) - \mathbf{x}(u, v_1)}{v - v_1}, \frac{\mathbf{y}(r, s) - \mathbf{y}(r, s_1)}{s - s_1} \right). \tag{9.8}$$

The determinant b is a polynomial and its monomial support with respect to (v, s) is $\mathcal{S} = \{1, v, s, vs\}$ and similarly for (v_1, s_1), where $\mathcal{S}_1 = \{1, v_1, s_1, v_1 s_1\}$. So, a monomial of b is a product of an element of $\mathcal{S}$ and of an element of $\mathcal{S}_1$. Then, we form the 4×4 matrix whose entries are the coefficients of b indexed by the product of the two sets $\mathcal{S}$ and $\mathcal{S}_1$. This matrix contains only the variables u and r and is called the Bezoutian matrix. In our case, its determinant is a polynomial in (u, r) equal to the desired resultant $R(u, r)$ (deg(R)=24 and $\deg_u(R)$=$\deg_r(R)$=16) and it gives an implicit curve which corresponds to the projection of $\mathcal{C}$ in the (u, r) space.

Then, we analyse the topology of this curve (see [17] and [30]) and we trace it (see [16] and [20]). Finally, for each $(u_0, r_0) \in [0, 1]^2$ such that $R(u_0, r_0) = 0$, we can determine if there exists a pair $(v_0, s_0) \in [0, 1]^2$ such that $\mathbf{x}(u_0, v_0) = \mathbf{y}(r_0, s_0)$

(solve a polynomial system of three equations with two separated unknowns cf bidegree (2,2)) and thus we can avoid the problem of the phantom components (see Fig. 9.2). We lift the obtained points in the 3D space to give the intersection locus. Note that this method can also give the projection of C in the (v, s) space by the same kind of computation.

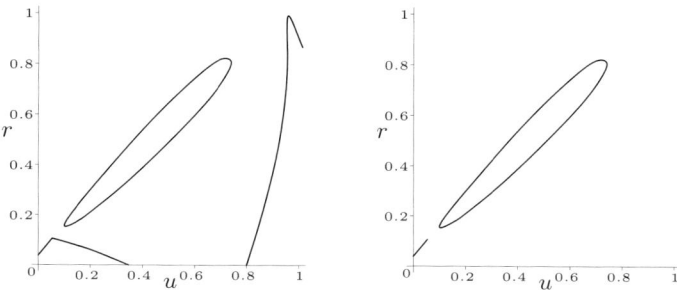

Fig. 9.2. Projection of C in the (u, r) space with (left) and without (right) phantom components. This curve corresponds to the example of Figure 9.6, page 175.

9.4 Approximate implicitization by a quartic surface

In this section, we apply the technique of approximate implicitization to compute the intersection of two biquadratic patches.

9.4.1 Approximate implicitization

The implicitization problem – which consists in finding an implicit equation (an algebraic representation) for a given parameterized rational surface – can be adressed by using several approaches, e.g., using resultants or Groebner bases [8, 9, 13]. However, the implicitization is very time consuming because of the degree of the implicit equation: for a generic parameterized surface of bidegree (n_1, n_2), the implicit equation has degree $2n_1n_2$. Also, all rational parametric curves and surfaces have an algebraic representation, but the reverse is not true; the relationship between the parametric and the algebraic representations can be very complex (problem of "phantom components"). Thus, we can try to find an algebraic approximation of a given parameterized surface for which the computation is more efficient and which contains less phantom components.

Consider a polynomial parameterized surface $\mathbf{x}(u, v)$ with the domain $[0, 1]^2$, and let d be a positive integer (the degree of the approximate implicit equation) and

$\epsilon \geq 0$ (the tolerance). Following [12], the approximate implicitization problem consists in finding a non–zero polynomial $P \in \mathbb{R}[x, y, z]$ of degree d such that

$$\forall (u, v) \in [0, 1]^2, P\left(\mathbf{x}(u, v) + \alpha(u, v)\, \mathbf{g}(u, v)\right) = 0 \qquad (9.9)$$

with $|\alpha(u, v)| \leq \epsilon$ and $||\mathbf{g}(u, v)||_2 = 1$. Here, α is the error function and $\mathbf{g}$ is the direction for error measurement, e.g., the unit normal direction of the surface patch.

9.4.2 Approximate implicitization of a biquadratic surface

The main question of the approximate implicitization problem is how to choose the degree. A key ingredient for this choice seems to be the topology, especially if the initial surface has self–intersections. The use of degree 4 was suggested by Tor Dokken; after several experiments he concluded that the algebraic surfaces of degree 4 provide sufficiently many degrees of freedom to approximate most cases encountered in practice. In the case of a biquadratic surface, where the exact implicit equation has degree 8, using degree 4 seems to be a reasonable trade-off.

We describe two methods for approximate implicitization by a quartic for a biquadratic surface. The approximate implicit equation is

$$P(x, y, z) = \sum_{i=0}^{4} \sum_{j=0}^{4-i} \sum_{k=0}^{4-i-j} b_{ijk}\, x^i y^j z^k \qquad (9.10)$$

with the unknown coefficients $b = (b_{000}, b_{100}, \ldots, b_{004}) \in \mathbb{R}^{35}$. Let $\beta(u, v)$ be the vector formed by the tensor–product Bernstein polynomials of bidegree (8,8).

Dokken's method.

This method, which is described in more detail in [12], proceeds as follows:

1. Factorize $P(\mathbf{x}(u, v)) = (Db)^T \beta(u, v)$ where D is a 81×35 matrix.
2. Generate a singular values decomposition (SVD) of D.
3. Choose b as the vector corresponding to the smallest singular value of D.

Note that this method is general and does not use the fact that we have a biquadratic surface. Hereafter, we use an adapted method based on the geometry of the surface of bidegree (2,2). Also, the computation of the singular value decomposition needs floating point numbers.

Geometric method using evaluation:

This approach consists in constructing some pertinent geometrical constraints to give a linear system of equations (with the unknowns $b_{000}, b_{100}, \ldots, b_{004}$), and then solving the resulting system by a singular values decomposition. In our method, we characterize some conics, especially the four border conics and two interior conics:

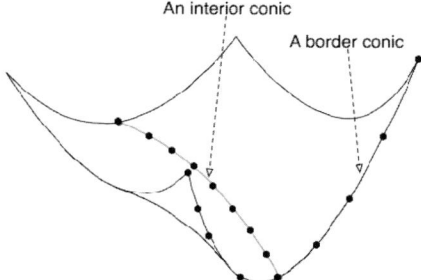

Fig. 9.3. Characterization of a conic in a biquadratic patch by 9 points

$$\begin{aligned}
C_1 &= \mathbf{x}([0,1] \times \{0\}), && C_2 = \mathbf{x}([0,1] \times \{1\}) \\
C_3 &= \mathbf{x}(\{0\} \times [0,1]), && C_4 = \mathbf{x}(\{1\} \times [0,1]) \\
C_5 &= \mathbf{x}(\{\tfrac{1}{2}\} \times [0,1]), && C_6 = \mathbf{x}([0,1] \times \{\tfrac{1}{2}\})
\end{aligned} \qquad (9.11)$$

Lemma 1. *If the quartic surface $\{P = 0\}$ contains 9 points of any of the 6 conics C_i, then $C_i \subset \{P = 0\}$, see Fig. 9.3.*

Proof. C_i is of degree 2 and P is of degree 4, so by Bézout's theorem, if there are more than 8 elements in $C_i \cap \{P = 0\}$, then $C_i \subset \{P = 0\}$.

Using this geometric observation, we construct a linear system and solve it approximately via SVD; this leads to an algebraic approximation of $\mathbf{x}(u, v)$ by a degree 4 surface.

9.4.3 Application to the intersection problem

In order to compute the intersection curves, we apply the approximate implicitization to one of the patches and compose it with the second one. This leads to an implicit representation of the intersection curve in one of the parameter domains, which can then be traced and analyzed using standard methods for planar algebraic curves.

These two approximate implicitization methods are very efficient and suitable for general cases, but the results are not always satisfactory. When the given biquadratic patch is simple (i.e. with a certain flatness and without singularity and self–intersection) the approximation is very close to the initial surface. So, to use this method for a general biquadratic surface, we combine it, if needed, with a subdivision method (Casteljau's algorithm). The advantage is twofold, we exclude domains without intersections (by using bounding boxes) and avoid some unwanted configurations with a curve of self-intersection (use Hohmeyer's criterion [19]). For more complicated singularities, the results are definitively not satisfactory.

Note that even if we have a good criterion in the subdivision step, we still may have problems with phantom components (but in general fewer), so we have to cut off the extraneous branches as in the resultant method. This has to be done carefully

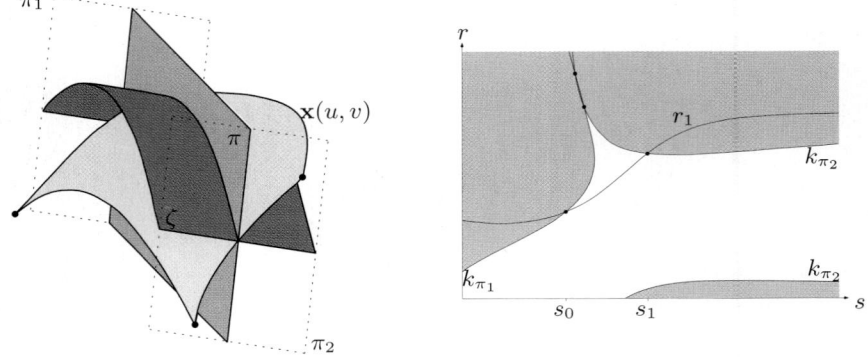

Fig. 9.4. Left: Representation of a parameter line as the intersection of a plane and a quadratic cylinder. Right: Identifying the intervals with feasible values of s.

in order to not discard points which do not correspond to phantom components. As another drawback – because of the various approximations – it is rather difficult to obtain certified points on the intersection locus. The use of approximate implicitization is clearly a numerical method, and it can give only approximate answers, even in the case of exact input.

9.5 Tracing intersections of parameter lines

In order to be able to trace the (self–) intersection curve(s), we have to find at least one point for each segment. We generate these points by intersecting the parameter lines of the first Bézier surface with the second one (see also [19]).

9.5.1 Intersection of a parameter line

A parameter line of $\mathbf{x}(u, v)$ for a constant rational value $u = u_0$ takes the form

$$\mathbf{p}(v) = \mathbf{x}(u_0, v) = \mathbf{a}_0(u_0) + \mathbf{a}_1(u_0)\, v + \mathbf{a}_2(u_0)\, v^2$$

with certain rational coefficient vectors $\mathbf{a}_i \in \mathbb{Q}^3$. It is a quadratic Bézier curve, hence we can represent it as the intersection of a *plane* and a *quadratic cylinder*, see Fig. 9.4, left. Since we are only interested in the intersection of these two surfaces in a certain region, we introduce two additional *bounding planes* π_1 and π_2. In the particular case that the parameter line is a straight line, we represent it as an intersection curve of two orthogonal planes.

In order to compute the intersection of the parameter line with the second surface patch $\mathbf{y}(r, s)$, we use the following algorithm.

1. Describe the parameter line as the intersection of a plane and a cylinder.

2. Intersect the plane with the second patch $\mathbf{y}(r, s)$ and compute the intersection $\mathcal{I}$.
3. Restrict the intersection curve(s) $\mathcal{I}$ to the region of interest.
4. Intersect the cylinder with the restricted intersection curve(s).

The four steps of the algorithm will now be explained in some more detail.

Defining the plane, the cylinder and the two bounding planes.

The parameter line and its three control points are coplanar. For computing the normal vector $\mathbf{n}$ of the plane, we have to evaluate the cross product of two difference vectors of the control points. The plane is given by the zero set of a linear polynomial

$$\pi(u_0)(x, y, z) = e_0(u_0) + n_1(u_0)\, x + n_2(u_0)\, y + n_3(u_0)\, z. \tag{9.12}$$

By extruding the parameter line in the direction of the normal vector of the plane, we obtain the parametric form of the quadratic cylinder, which intersects the plane orthogonally,

$$\mathbf{w}(u_0)(p, q) = \mathbf{x}(u_0, p) + q \cdot \mathbf{n}. \tag{9.13}$$

The implicitation of the cylinder is slightly more complicated. There exist two possibilities: we can either use Sylvester resultants or the method of comparing coefficients. In both cases we will get an equation of the form

$$\begin{aligned}
\zeta(u_0)(x, y, z) := {} & a_0(u_0) + a_1(u_0)\, x + a_2(u_0)\, y + a_3(u_0)\, z \\
& + a_4(u_0)\, x\, y + a_5(u_0)\, x\, z + a_6(u_0)\, y\, z \\
& + a_7(u_0)\, x^2 + a_8(u_0)\, y^2 + a_9(u_0)\, z^2 = 0. \tag{9.14}
\end{aligned}$$

Now we have both the plane and the cylinder in their implicit representation. Note that this is a semi-implicit representation in the sense of [6].

If the parameter line degenerates into a straight line, then we choose two planes through it which intersect orthogonally. Note that we use exact rational arithmetic, in order to avoid any robustness problems.

Finally, we create the two planes $\pi_1(x, y, z)$ and $\pi_2(x, y, z)$ which bound the parameter line. For instance, one may choose the two normal planes of the parameter line at its boundary points; this choice is always possible, provided that the curve segment is not too long (which can be enforced by using subdivision). Alternatively one may use the planes spanned by the boundary curves, but these planes may have an additional intersection with the parameter line in the region of interest.

Intersection of the plane and the second patch $\mathbf{y}(r, s)$.

Substituting the second Bézier surface $\mathbf{y}(r, s)$ into the equation (9.12) of the plane leads to a biquadratic equation in r and s. We can treat it as a quadratic polynomial in r with coefficients depending on s.

$$\pi(\mathbf{y}(r, s)) = a(s)\, r^2 + b(s)\, r + c(s) = 0. \tag{9.15}$$

For each value of s, we obtain two solutions $r_1(s)$ and $r_2(s)$ of the form

$$r_{1,2}(s) = -\frac{b(s)}{2\,a(s)} \pm \sqrt{d(s)} \quad \text{with} \quad d(s) = \frac{b(s)^2}{4\,a(s)^2} - \frac{c(s)}{a(s)}. \tag{9.16}$$

These solutions parameterize the two branches of the intersection curve $\mathcal{I}$ in the rs–parameter domain of the second patch. By solving several quadratic equations we determine the intervals $S_{i,j} \subset [0,1]$, where $d(s) \geq 0$ and $0 \leq r_i(s) \leq 1$ holds; this leads to a (list of) feasible domain(s) (i.e., intervals) for each branch of the intersection curve.

By composing (9.16) with $\mathbf{y}$ we obtain the two branches $\mathbf{k}_1(s)$ and $\mathbf{k}_2(s)$ of the intersection curve $\mathcal{I}$,

$$\mathbf{k}_{1,2}(s) = \mathbf{y}(r_{1,2}(s), s) = \frac{1}{a(s)^2}\mathbf{h}(s) \pm \frac{\sqrt{d(s)}}{a(s)}\mathbf{l}(s) + d(s)\mathbf{m}(s) \tag{9.17}$$

where the components of $\mathbf{h}(s)$, $\mathbf{l}(s)$ and $\mathbf{m}(s)$ are polynomials of degree 6, 4, and 2, respectively.

Restriction to the region of interest.

Since the region of interest is located between the planes $\pi_1(x, y, z)$ and $\pi_2(x, y, z)$, the two inequalities

$$\pi_1(x, y, z) \geq 0 \quad \text{and} \quad \pi_2(x, y, z) \leq 0 \tag{9.18}$$

have to be satisfied. By intersecting each bounding plane with the second Bézier surface $\mathbf{y}(r, s)$ in a similar way as described for $\pi(u_0)(x, y, z)$, we obtain

$$k_{\pi_1}(s) := \pi_1(\mathbf{y}(r(s), s)) \geq 0 \quad \text{and} \quad k_{\pi_2}(s) := \pi_2(\mathbf{y}(r(s), s)) \leq 0 \tag{9.19}$$

This leads to additional constraints for the feasible values of the parameter s. For each branch of the intersection curve we create the (list of) feasible domain(s) and store it. The bounds of the intervals can be computed by solving three systems of two biquadratic equations or – equivalently – by solving a system of three polynomials of degree 8, which are obtained after eliminating the parameter r. Here, we represent the polynomials in Bernstein–Bézier form and use a Bézier–clipping–type technique see [14, 25, 26, 28], applied to floating point numbers.

Example 2. For a parameter line $u = u_0$ of two biquadratic Bézier surface patches $\mathbf{x}(u, v)$ and $\mathbf{y}(r, s)$, Fig. 9.4, right, shows the rs–parameter domain of the second patch. Only the first branch $r_1(s)$ of the intersection curve is present. The bounds $0 \leq r \leq 1$ do not impose additional bounds on s in this case. However, the intersection with the bounding planes π_1 and π_2 produces two additional curves, which have to be intersected with the curve $s = r_1(s)$, leading to two bounds s_0 and s_1 of the feasible domain.

Intersection of the cylinder and the intersection curves.

We substitute the parametric representation of the intersection curve into the implicit equation (9.14) of the cylinder and obtain

$$\zeta(u_0)(s) = p_1(s) + p_2(s)\sqrt{d(s)} + p_3(s)\left(\sqrt{d(s)}\right)^2 +$$
$$+ p_4(s)\left(\sqrt{d(s)}\right)^3 + p_5(s)\left(\sqrt{d(s)}\right)^4 = 0 \qquad (9.20)$$

where the polynomials $p_j(s)$ are of degree 12. In order to eliminate the square root, we use the following trick. We split $\zeta(s, d(s)) = A - B$, where A and B contain all even and odd powers of $\sqrt{d}$, respectively. The equation $A - B = 0$ is then replaced with $A^2 \cdot d(s) - (B \cdot \sqrt{d(s)})^2 = 0$. This leads to a polynomial of degree 24 in one variable. After factoring out the discriminant, we obtain a polynomial of degree 16 in s. Note that this agrees with the theoretical number of intersections of a biquadratic surface, which has algebraic order 8, with a quadratic curve.

Finally, we solve this polynomial within all the feasible intervals of s, which were detected in the previous steps. Until this point we used symbolic computations. Now – after generating the Bernstein–Bézier representation – we change to floating-point numbers and use a Bézier–clipping–type method to find all roots within the feasible domain(s). These roots correspond to intersection points of the parameter line of the first patch with the second patch.

9.5.2 Global structure of the intersection curve

For each value $u = u_0$, the parameter line $\mathbf{x}(u_0, v)$ has a certain number of intersection points with the second patch. If u_0 varies continuously, then the number of intersection points may change only if

(1) one of the intersection points is at the boundary of one of the patches (boundary points) or
(2) the parameter line of the first patch touches the second patch (turning points) .

The algorithm for analyzing the global structure of the intersection curve proceeds in two steps: First we detect those values of u_0 where the number of intersection points changes, and order them. This leads to a sequence of critical u_0- values,

$$0 = u_0^{(0)} < u_0^{(1)} < \ldots < 1 = u_0^{(K)}. \qquad (9.21)$$

In the second step, we analyze the intersection of the parameter lines $u_0 = (u_0^{(i)} + u_0^{(i+1)})/2$ with the second patch. Since the number of intersection points between any two critical values remains constant, we can now either trace the segment using conventional techniques for tracing surface–surface intersections (see [20]) or generate more points by analyzing more intersections with parameter lines.

In the remainder of this section we address the computation of the critical u_0 values.

Boundary points.

Such points correspond to intersections of the boundary parameter lines of one surface with the other one. In order to compute them, we apply the algorithm for intersecting parameter lines with a biquadratic patch to the $2 \cdot 4$ boundary parameter lines of the two surfaces.

Turning points.

We consider the turning points of $\mathbf{x}(u, v)$ in respect to u. Let $\mathbf{y}_r$ and $\mathbf{y}_s$ denote the partial derivatives of $\mathbf{y}(r, s)$. Several possibilities for computing the turning points exist.

1. The two surfaces $\mathbf{x}(u_0, v)$ and $\mathbf{y}(r, s)$ intersect, $\mathbf{x}(u_0, v) = \mathbf{y}(r, s)$, and the tangent vector of the parameter line lies in the tangent plane of the second patch,

$$\mathbf{x}_u \cdot (\mathbf{y}_r \times \mathbf{y}_s) = 0. \tag{9.22}$$

   These conditions lead to a system of four polynomial equations for four unknowns, which has to be solved for u.
2. By using the previous geometric result, we may eliminate the variable v, as follows. First, the plane spanned by the parameter line has to contain the point $\mathbf{y}(r, s)$,

$$\pi(u_0)(\mathbf{y}(r, s)) = 0, \tag{9.23}$$

which gives an equation of degree $(6, 2, 2)$ in (u_0, r, s). Second, the cylinder has to contain the point,

$$\zeta(u_0)(\mathbf{y}(r, s)) = 0, \tag{9.24}$$

which leads to an equation of degree $(16, 4, 4)$. Finally, the tangent vector of the parameter line has to be contained in the tangent plane of the second patch. Since the tangent of the parameter line is parallel to the cross product of the gradient of the plane and the gradient of the cylinder, the third condition gives an equation of degree $(18, 5, 5)$,

$$\det \left[\mathbf{y}_r, \ \mathbf{y}_s, \ \nabla \pi(u_0)(\mathbf{y}(r, s)) \times \nabla \zeta(u_0)(\mathbf{y}(r, s)) \right] = 0. \tag{9.25}$$

For solving either of these two systems of polynomial equations, we use again a Bézier–clipping–type algorithm [14, 25, 28].

9.6 Self–intersections of biquadratic surface patches

In order to detect the self–intersection curves of any of the two patches, the methods for surface–surface intersections have to be modified. The computation of the self–intersection locus by using approximate implicitization is not discussed here, since it was already treated in [31]. Instead we focus on the other two techniques.

9.6.1 Resultant-based method

In the parameter domain $[0, 1]^4$, the self–intersection curve of the first patch forms the set

$$\left\{ (u_1, v_1, u_2, v_2) \in [0, 1]^4 \mid (u_1, v_1) \neq (u_2, v_2) \text{ and } \mathbf{x}(u_1, v_1) = \mathbf{x}(u_2, v_2) \right\}.$$
(9.26)

This locus is the real trace of a complex curve. We assume that it is either empty or of dimension 0 or 1. We do not consider degenerate cases, such as a plane which is covered twice. In the examples presented below (see Section 9.7), the self–intersection locus is a curve in $\mathbb{R}^4$.

We use the following change of coordinates to discard the unwanted trivial component $(u_1, v_1) = (u_2, v_2)$. Let (u_2, v_1) be a pair of parameters in $[0, 1]^2$, $(l, k) \in \mathbb{R}^2$ and let $u_1 = u_2 + l$, $v_2 = v_1 + lk$. If we suppose that we have $(u_1, v_1) \neq (u_2, v_2)$, then $l \neq 0$. Hence $\mathbf{x}(u_1, v_1) = \mathbf{x}(u_2, v_2)$ if and only if $\mathbf{x}(u_2+l, v_1) = \mathbf{x}(u_2, v_1+lk)$. We suppose now that (u_2, v_1, l, k) verifies this last relation.

Let $\tilde{T}(u_2, v_1, l, k)$ be the polynomial $\frac{1}{l} \left[\mathbf{x}(u_2 + l, v_1) - \mathbf{x}(u_2, v_2 + lk) \right]$, its degree in (u_2, v_1, l, k) is $(2, 2, 1, 2)$ and the monomial support with respect to (l, k) contains only $k^2 l, k, l$ and 1. We can decrease the degree by introducing

$$T(u_2, v_1, m, k) = m\tilde{T}(u_2, v_1, \frac{1}{m}, k).$$
(9.27)

Then in $T(u_2, v_1, m, k)$, the monomial support in (m, k) consists only of $1, m, k^2$ and km. So, we can write T in a matrix form:

$$T(u_2, v_1, m, k) = \begin{pmatrix} a_1(u_2, v_1) \ b_1(u_2, v_1) \ c_1(u_2, v_1) \ d_1(u_2, v_1) \\ a_2(u_2, v_1) \ b_2(u_2, v_1) \ c_2(u_2, v_1) \ d_2(u_2, v_1) \\ a_3(u_2, v_1) \ b_3(u_2, v_1) \ c_3(u_2, v_1) \ d_3(u_2, v_1) \end{pmatrix} \begin{pmatrix} 1 \\ m \\ k^2 \\ km \end{pmatrix}$$
(9.28)

By Cramer's rule, we get

$$m = \frac{D_2}{D_1}, \quad k^2 = \frac{D_3}{D_1}, \quad \text{and} \quad km = \frac{D_4}{D_1}$$
(9.29)

with

$$D_1 = \begin{vmatrix} b_1 & c_1 & d_1 \\ b_2 & c_2 & d_2 \\ b_3 & c_3 & d_3 \end{vmatrix}, \ D_2 = \begin{vmatrix} -a_1 & c_1 & d_1 \\ -a_2 & c_2 & d_2 \\ -a_3 & c_3 & d_3 \end{vmatrix}, \ D_3 = \begin{vmatrix} b_1 & -a_1 & d_1 \\ b_2 & -a_2 & d_2 \\ b_3 & -a_3 & d_3 \end{vmatrix}, \ D_4 = \begin{vmatrix} b_1 & c_1 & -a_1 \\ b_2 & c_2 & -a_2 \\ b_3 & c_3 & -a_3 \end{vmatrix}.$$

Let $Q(u_2, v_1)$ be the polynomial $Q = D_4^2 D_1 - D_2^2 D_3$.

Lemma 3. *The implicitly defined curve* $\left\{ (u_2, v_1) \in [0, 1]^2 \mid Q(u_2, v_1) = 0 \right\}$ *is the projection of the self–intersection locus (given by the set (9.26) but in $\mathbb{C}^4$) into the parameters domain* $(u_2, v_1) \in [0, 1]^2$.

Fig. 9.5. A self–intersection of a surface with a cuspidal point

Proof. $Q(u_2, v_1) = 0$ is the only algebraic relation (of minimal degree) between u_2 and v_1 such that

$$\forall (u_2, v_1) \in [0, 1]^2, Q(u_2, v_1) = 0 \Rightarrow \exists (m, k) \in \mathbb{C}^2, T(u_2, v_1, m, k) = 0.$$

This lemma provides a method to compute the self–intersection locus, we just have to trace the implicit curve $Q(u_2, v_1) = 0$ and for every point (u_2, v_1) on this curve, we obtain by continuation the corresponding point $(u_1, v_2) \in [0, 1]^2$ if it exists (see the results on Fig. 9.9). So it suffices to characterize the bounds of these segments of curves.

9.6.2 Parameter-line-based method

For computing the self–intersection curves, we use the same algorithm as described in Section 9.5. We intersect the surface $\mathbf{x}(u_0, v)$ with itself $\mathbf{x}(r, s)$. In this case, both the "plane" equation (9.23) and the "cylinder" equation (9.24) contain the linear factor $(r - u_0)$, which has to be factored out. The computation of turning points as in section 9.5.2 leads us to two different types: the usual ones and cuspidal points (see Fig. 9.5).

9.7 Examples

The three methods presented in this paper (using resultants, via approximate implicitization, and by analyzing the intersections with parameter lines) work well for most standard situations usually encountered in practice. In this section, we present three representative examples. Additional ones are available at [21].

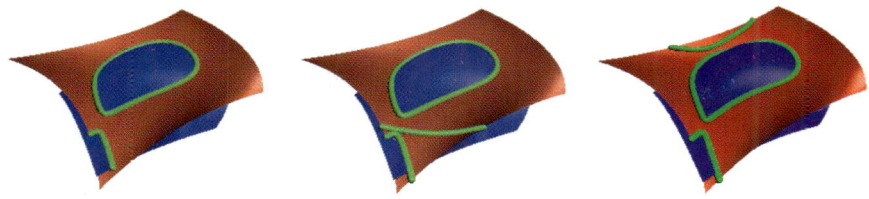

Fig. 9.6. First example. Left and center: Result of the resultant method after and before eliminating phantom branches. Right: result of the approach using approximate implicitization.

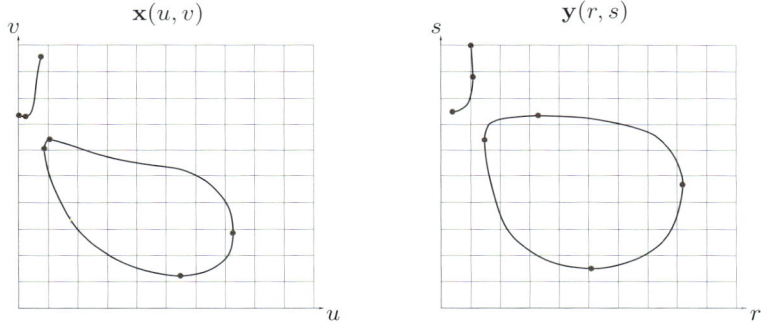

Fig. 9.7. First example. The intersection curves in the parameter domains of both surface patches, generated by the parameter–line based technique. Boundary points and turning points have been marked by grey circles.

9.7.1 First example

We consider two biquadratic surfaces with an open and a closed component of the intersection curve. The two surfaces have the control points

$$
\underbrace{\begin{bmatrix} \left(\frac{1}{7},0,\frac{3}{5}\right) & \left(\frac{3}{5},\frac{1}{5},\frac{3}{4}\right) & \left(1,0,\frac{7}{10}\right) \\ \left(\frac{3}{8},\frac{4}{9},\frac{2}{3}\right) & \left(\frac{2}{3},\frac{3}{4},\frac{1}{3}\right) & \left(\frac{6}{7},\frac{3}{8},\frac{5}{7}\right) \\ \left(\frac{1}{5},\frac{6}{7},\frac{4}{7}\right) & \left(\frac{3}{4},\frac{7}{8},\frac{3}{4}\right) & \left(\frac{7}{8},\frac{7}{9},\frac{5}{8}\right) \end{bmatrix}}_{\mathbf{x}(u,v)} \quad \text{and} \quad \underbrace{\begin{bmatrix} \left(\frac{2}{7},\frac{1}{7},\frac{2}{5}\right) & \left(\frac{3}{5},\frac{1}{10},\frac{2}{3}\right) & \left(1,0,\frac{4}{5}\right) \\ \left(\frac{3}{8},\frac{4}{9},\frac{2}{3}\right) & \left(\frac{1}{3},\frac{1}{2},1\right) & \left(\frac{5}{7},\frac{3}{8},\frac{2}{7}\right) \\ \left(\frac{1}{5},\frac{6}{7},\frac{3}{7}\right) & \left(\frac{3}{4},\frac{7}{8},\frac{5}{8}\right) & \left(\frac{7}{8},\frac{4}{7},\frac{1}{2}\right) \end{bmatrix}}_{\mathbf{y}(r,s)}.
$$

By using the *resultant method*, a phantom component appears (see Fig. 9.6, center). It can be cut off as described in Section 9.3.2 (see Fig. 9.6, left).

Similar to the resultant method, the *approximate implicitization* produces a phantom component (see Fig. 9.6, right). However, when we cut it off, we obtain only very few certified points on the intersection locus as described in section 9.4.3.

The *parameter-line-based approach* finds both parts of the intersection curve, but no phantom components. One segment is closed and has two turning points with respect to each parameter u, v, r and s. The other segment has two boundary points

$u = 0$ and $s = 1$ and also possesses a turning point with respect to v and another one with respect to r (see Fig. 9.7).

9.7.2 Second example

The control points of the two biquadratic surfaces

$$\left[\begin{array}{ccc} \left(\frac{501}{775}, \frac{388}{775}, \frac{588}{775}\right) & \left(\frac{347}{775}, \frac{276}{775}, \frac{479}{775}\right) & \left(\frac{309}{775}, \frac{604}{775}, \frac{498}{775}\right) \\ \left(\frac{553}{775}, \frac{454}{775}, \frac{293}{775}\right) & \left(\frac{336}{775}, \frac{382}{775}, \frac{469}{775}\right) & \left(1, \frac{426}{775}, \frac{137}{775}\right) \\ \left(\frac{337}{775}, \frac{308}{775}, \frac{258}{775}\right) & \left(\frac{517}{775}, 0, \frac{367}{775}\right) & \left(\frac{533}{775}, \frac{492}{775}, \frac{564}{775}\right) \end{array}\right]$$

$$\underbrace{}_{\mathbf{x}(u,v)}$$

$$\left[\begin{array}{ccc} \left(\frac{492}{775}, \frac{67}{155}, \frac{522}{775}\right) & \left(\frac{543}{775}, \frac{322}{775}, \frac{117}{775}\right) & \left(\frac{346}{775}, \frac{13}{155}, \frac{4}{5}\right) \\ \left(\frac{113}{155}, \frac{392}{775}, \frac{58}{155}\right) & \left(\frac{632}{775}, \frac{469}{775}, \frac{413}{775}\right) & \left(\frac{307}{775}, \frac{514}{775}, \frac{564}{775}\right) \\ \left(\frac{602}{775}, \frac{129}{775}, \frac{274}{775}\right) & \left(\frac{669}{775}, \frac{692}{775}, \frac{53}{155}\right) & \left(\frac{488}{775}, \frac{219}{775}, \frac{412}{775}\right) \end{array}\right]$$

$$\underbrace{}_{\mathbf{y}(r,s)}$$

were generated by using a pseudo–random number generator.

The *resultant–based technique* leads to several phantom components (see Fig. 9.8, center), which can be cut off as described previously (see Fig. 9.8, left).

The combined use of *subdivision* and *approximate implicitization* produces even more phantom components (see Fig. 9.8, right). This is due to the fact that the subdivision generates more implicitly defined surfaces. Eventually we obtain sufficiently many points to draw the correct intersection curves.

We also computed the self–intersection curve (see Fig. 9.9) with the help of the method described in Section 9.6.1.

When using the *parameter–line based approach*, this example does not lead to any difficulties. The intersection curve consists of three segments (see Fig. 9.10). The first Bézier surface patch $\mathbf{x}(u, v)$ has one self–intersection curve, while the second one $\mathbf{y}(r, s)$ intersects itself three times and has two cuspidal points.

9.7.3 Third example

The two biquadratic surface patches with the control points

$$\left[\begin{array}{ccc} \left(0, \frac{1}{7}, \frac{4}{5}\right) & \left(\frac{3}{5}, \frac{1}{13}, \frac{1}{3}\right) & \left(1, 0, \frac{4}{5}\right) \\ \left(\frac{1}{8}, \frac{4}{9}, \frac{11}{40}\right) & \left(\frac{1}{3}, \frac{34}{65}, \frac{3}{4}\right) & \left(\frac{6}{7}, \frac{3}{8}, -\frac{16}{35}\right) \\ \left(\frac{1}{5}, \frac{6}{7}, \frac{4}{5}\right) & \left(\frac{3}{4}, \frac{443}{520}, \frac{3}{8}\right) & \left(\frac{7}{8}, 1, \frac{14}{15}\right) \end{array}\right] \quad \text{and} \quad \left[\begin{array}{ccc} \left(0, \frac{1}{7}, \frac{1}{5}\right) & \left(\frac{3}{5}, \frac{1}{10}, \frac{1}{3}\right) & \left(1, 0, \frac{1}{5}\right) \\ \left(\frac{1}{8}, \frac{4}{9}, \frac{7}{8}\right) & \left(\frac{1}{3}, \frac{1}{2}, \frac{3}{4}\right) & \left(\frac{6}{7}, \frac{3}{8}, \frac{1}{7}\right) \\ \left(\frac{1}{5}, \frac{6}{7}, \frac{1}{5}\right) & \left(\frac{3}{4}, \frac{7}{8}, \frac{3}{8}\right) & \left(\frac{7}{8}, 1, \frac{1}{3}\right) \end{array}\right]$$

$$\underbrace{}_{\mathbf{x}(u,v)} \qquad\qquad \underbrace{}_{\mathbf{y}(r,s)}$$

touch each other along a parameter line.

The *resultant-based approach* leads to an implicitly defined curve which describes the intersection. Due to the special situation, it contains the square of this

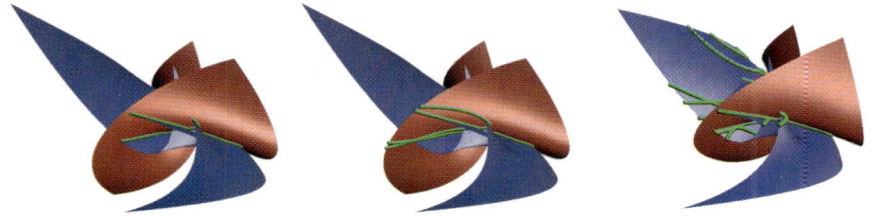

Fig. 9.8. Second example: Left and center: result of the resultant method after and before eliminating phantom branches. Right: result obtained by using approximate implicitization.

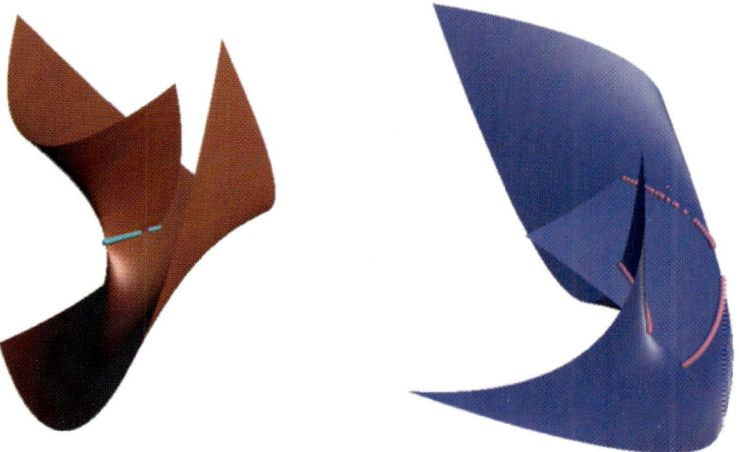

Fig. 9.9. Second example: Self intersections, computed with the method described in Section 9.6.1.

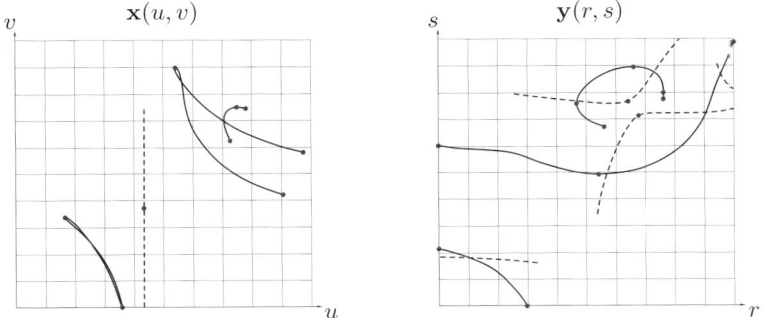

Fig. 9.10. Second example: Intersection (solid, black) and self–intersection (dashed, grey) curves in the parameter domains of both surface patches, generated by the parameter–line based technique. Boundary points and turning points have been marked by grey circles.

equation. A direct tracing of the curve is difficult, since the use of a standard predictor/corrector method is numerically unstable. However, one may factorize the equation, and apply the tracing to the individual factors without probems. This leads to the curve shown in Fig. 9.11, left.

The technique of *approximate implicitization* is not well suited to deal with this very specific situation: the approximation produces either an empty intersection or two curves which are close to each other (see Fig. 9.11, right).

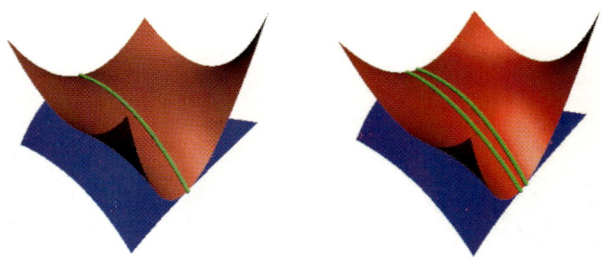

Fig. 9.11. Third example. Left: Result of the resultant method and of the parameter–line based approach. Right: result of the use of approximate implicitization.

The *parameter-line-based approach* finds two boundary points and it produces – for each value of $u = u_0$ – the correct intersection point of the parameter line with the other patch. The convergence of the Bézier clipping slows down to a linear rate, due to the presence of a double root. Also, it is difficult to trace the intersection curve by using a geometric predictor/corrector technique. Instead, we computed the intersection points for many values of u_0 and arrived at a result which is very similar to Fig. 9.11, left.

9.8 Conclusion

We presented three different algorithms for computing the intersection and self–intersection curves of two biquadratic Bézier surface patches. We implemented the methods and applied them to many test cases. Three of them have been presented in this paper.

The *resultant–based technique* was able to deal with all test cases. It may produce additional 'phantom' branches, which have to be eliminated by carefully analyzing the result of the elimination. As an advantage, one may – in the case of two surface patches that touch each other – factorize the implicit equation of the intersection curve, in order to obtain a stable representation, which can then be traced robustly.

After experimenting with *approximate implicitization* we arrived at the conclusion that this method is not to be recommended for biquadratic patches. On the one hand, it is not suited for avoiding problems with phantom branches. On the other hand, the use of an approximate technique introduces inaccuracies, which may cause problems with singular and almost singular situations. We feel that this price for using a lower degree implicit representation is too high.

The *parameter–line based approach* adds some geometric interpretations to the process of eliminating variables from the problem. As an advantage, it is possible to correctly establish the region(s) of interest. This avoids problems with unwanted branches of the (self–) intersection curves. In the case of two touching surfaces, using this approach becomes more expensive, since standard techniques for tracing the intersection cannot be applied.

Acknowledgment

This research was supported by the European Union through project IST 2001–35512 'Intersection algorithms for geometry based IT applications using approximate algebraic methods' (GAIA II), by the Austrian Science Fund through the Joint Research Programme FSP S92 'Industrial Geometry', and by Aim@Shape (IST NoE 506766).

References

1. L. Andersson, J. Peters, and N. Stewart, Self-intersection of composite curves and surfaces, *Computer Aided Geometric Design, 15 (1998)*, pp. 507–527.
2. L. Busé, Etude du résultant sur une variété algébrique, *PhD thesis, University of Nice*, December 2001.
3. L. Busé and C. D'Andrea, Inversion of parameterized hypersurfaces by means of subresultants, *Proceedings ACM of the ISSAC 2004*, pp. 65–71.
4. L. Busé, M. Elkadi, and B. Mourrain, Using projection operators in Computer Aided Geometric Design, *In Topics in Algebraic Geometry and Geometric Modeling*, pp. 321–342, Contemporary Mathematics, AMS, 2003.
5. L. Busé, I.Z. Emiris and B. Mourrain, *MULTIRES*, http://www-sop.inria.fr/galaad/logiciels/multires.
6. L. Busé and A. Galligo, Using semi-implicit representation of algebraic surfaces, *Proceedings of the SMI 2004 conference, IEEE Computer Society*, pp. 342–345.
7. E.W. Chionh and R.N. Goldman, Using multivariate resultants to find the implicit equation of a rational surface, *The Visual Computer 8 (1992)*, pp. 171–180.
8. D. Cox, J. Little and D. O'Shea, Ideals, Varieties and Algorithms, *Springer-Verlag*, New York, 1992 and 1997.
9. D. Cox, J. Little and D. O'Shea, Using Algebraic Geometry, *Springer-Verlag*, New York, 1998.
10. C. D'Andrea, Macaulay style formulas for sparse resultants, *Trans. Amer. Math. Soc., 354(7) (2002)*, pp. 2595–2629.
11. T. Dokken, Aspects of Intersection Algorithms and Approximation, *Thesis for the doctor philosophias degree*, University of Oslo, Norway 1997.
12. T. Dokken, Approximate implicitization, *Mathematical Methods for Curves and Surfaces*, T. Lyche and L.L. Schumaker (eds.), Vanderbilt University Press, 2001, pp. 81–102.

13. T. Dokken and J.B. Thomassen, Overview of Approximate Implicitization, *Topics in Algebraic Geometry and Geometric modeling*, ed. Ron Goldman and Rimvydas Krasauskas, AMS series on Contemporary Mathematics CONM 334, 2003, pp. 169–184.
14. G. Elber and M-S. Kim, Geometric Constraint Solver using Multivariate Rational Spline Functions, *The Sixth ACM/IEEE Symposium on Solid Modeling and Applications, 2001*, pp. 1–10.
15. M. Elkadi and B. Mourrain, Some applications of Bezoutians in Effective Algebraic Geometry, *Rapport de Recherche 3572*, INRIA, Sophia Antipolis, 1998.
16. G. Farin, J. Hoschek and M-S. Kim, Handbook of Computer Aided Geometric Design, *Elsevier*, 2002.
17. L. González-Vega and I. Necula, Efficient topology determination of implicitly defined algebraic plane curves, *Comput. Aided Geom. Design, 19(9) (2002)*, pp. 719–743.
18. C. M. Hoffmann, Implicit Curves and Surfaces in CAGD, *Comp. Graphics and Appl. (1993)*, pp. 79–88.
19. M. E. Hohmeyer, A Surface Intersection Algorithm Based on Loop Detection, *ACM Symposium on Solid Modeling Foundations and CAD/CAM Applications, 1991*, pp. 197–207.
20. J. Hoschek and D. Lasser, Fundamentals of Computer Aided Geometric Design, *A.K. Peters*, 1993.
21. S. Chau and M. Oberneder, http://www.ag.jku.at/~margot/biquad
22. A. Khetan, The resultant of an unmixed bivariate system, *J. of Symbolic Computation, 36 (2003)*, pp. 425–442. http://www.math.umass.edu/~khetan/software.html
23. S. Krishnan and D. Manocha, An Efficient Surface Intersection Algorithm Based on Lower-Dimensional Formulation, *ACM Transactions on Graphics, 16(1) (1997)*, pp. 74–106.
24. L. Kunwoo, Principles of CAD/CAM/CAE Systems, *Addison-Wesley*, 1999.
25. B. Mourrain and J.-P. Pavone, Subdivision methods for solving polynomial equations, *Technical Report 5658*, INRIA Sophia-Antipolis, 2005.
26. T. Nishita, T.W. Sederberg and M. Kakimoto, Ray tracing trimmed rational surface patches, *Siggraph, 1990*, pp. 337–345.
27. N.M. Patrikalakis, Surface-to-surface intersections, *IEEE Computer Graphics and Applications, 13(1) (1993)*, pp. 89–95.
28. N. Patrikalakis and T. Maekawa, Chapter 25: Intersection problems, Handbook of Computer Aided Geometric Design (G. Farin and J. Hoschek and M.-S. Kim, eds.), *Elsevier*, 2002.
29. J.-P. Pavone, Auto-intersection des surfaces paramétrées réelles, *Thèse d'informatique de l'Université de Nice Sophia-Antipolis*, Décembre 2004.
30. J.P. Técourt, Sur le calcul effectif de la topologie de courbes et surfaces implicites, *PhD thesis in Computer Science at INRIA Sophia-Antipolis*, Décembre 2005.
31. J.B. Thomassen, Self-Intersection Problems and Approximate Implicitization, *Computational Methods for Algebraic Spline Surfaces, Springer*, pp. 155–170, 2005.
32. A. Vlachos, J. Peters, C. Boyd and J. L. Mitchell, Curved PN Triangles, Symposium on Interactive 3D Graphics, Bi-Annual Conference Series, ACM Press, 2001, 159–166.

10

Cube Decompositions by Eigenvectors of Quadratic Multivariate Splines

Ioannis Ivrissimtzis[1] and Hans-Peter Seidel[2]

[1] Durham University, UK
   ioannis.ivrissimtzis@durham.ac.uk
[2] MPI-Informatik, Saarbruecken, Germany
   hpseidel@mpi-inf.mpg.de

Summary. A matrix is called G-circulant if its columns and rows are indexed by the elements of a group G. When G is cyclic we obtain the usual circulant matrices, which appear in the study of linear transformations of polygons. In this paper, we study linear transformations of cubes and prisms using G-circulant matrices, where G is the direct product of cyclic groups. As application, we study the evolution of a single cell of an n-dimensional grid under the subdivision algorithm of the multivariate quadratic B-spline. Regarding the prism, we study its evolution under a tensor extension of the Doo-Sabin subdivision scheme.

10.1 Introduction

Knot insertion, refinement and subdivision algorithms not only make splines a very practical design tool but they also offer an insight into their mathematical properties. In fact, such algorithms can even be seen as alternative definitions of the splines. The latter has been proved a very fruitful approach leading to many generalizations, most notably the *subdivision surfaces*, which generalize bivariate B-splines over polygonal control meshes with arbitrary connectivity.

A subdivision surface is defined by an initial coarse polygonal mesh and a subdivision rule which refines the mesh by adding new vertices and connecting them with edges and faces creating a new denser mesh. Repetitive iterations of this process yield in the limit a smooth surface, which depends on the initial coarse mesh and the subdivision rule only.

Subdivision surfaces are studied locally, usually in the neighborhood of a vertex. At each subdivision step the connectivity of the subdivision mesh becomes larger, but, apart possibly from some initial steps, the local connectivity around the vertex we study does not change. Instead, after a subdivision step the same local connectivity corresponds to a smaller neighborhood of the vertex.

After each subdivision step the positions of the new vertices are given as linear combinations of the old vertices, usually described in the form of a *subdivision matrix*. The limit positions of the neighborhoods vertices are given by the limit of the mth power of the subdivision matrix, as m tends to infinity. That means that the properties of the subdivision surface are related to the spectral properties of the subdivision matrix.

The study of a subdivision surface through the study of the eigenstructure of the subdivision matrix can be facilitated by exploiting patterns of the matrix arising from the symmetries of the local connectivity. In the simplest example, the 1-ring neighborhood of a triangle mesh vertex has rotational symmetry of order k, where k is the valence of that vertex. This rotational symmetry is reflected in the rules of any subdivision scheme with reasonable behavior, resulting in subdivision matrices with circulant blocks.

Circulant matrices are simple patterned matrices with the property that each row is the previous row cycled forward one step. *Block-circulant* matrices are a generalization of the circulant. They have the same cyclic pattern, but this time on matrix blocks instead of single elements. An equivalent generalization is the *circulant-block* matrix, which has circulant matrices as blocks. *G-circulant* matrices generalize the circulant matrices into another direction. A matrix is called G-circulant if its columns and rows are indexed by the elements of a group G. The simplest cases of G-circulant matrices are those corresponding to abelian groups. As we will see below, such patterned matrices arise in volume subdivision processes, reflecting the symmetries of cubes and prisms.

10.1.1 Geometric decompositions to the eigenvectors of a patterned matrix

In this paper we study the geometric decompositions to the eigenvectors of patterned matrices. By this we mean decompositions into geometric objects that are only scaled by the patterned matrix. These geometric components can be seen as geometric interpretations of the eigenvectors of the patterned matrix, with the real eigenvectors corresponding to linear components and the complex eigenvectors giving planar components. By embedding these components into a higher dimensional space, we can use them to decompose higher dimensional objects, such as n-dimensional cubes and prisms. When there is no room for ambiguity, we will refer to the components as eigencomponents, or eigenvectors, or eigenpolygons, eigencubes and eigenprisms.

If a decomposition of a shape is given as

$$V_0 + V_1 + \cdots + V_{n-1} \tag{10.1}$$

and $\lambda_0, \lambda_1, \ldots, \lambda_{n-1}$ are the corresponding eigenvalues, then, after m consecutive transformations by the matrix the shape is given by

$$\lambda_0^m V_0 + \lambda_1^m V_1 + \cdots + \lambda_{n-1}^m V_{n-1} \tag{10.2}$$

We see that in the limit the shape is determined by the eigencomponents with the largest eigenvalues.

The use of the eigenvalues and the eigenvectors of the subdivision matrix, in a form similar to the one described by Eq. 10.2, is a standard tool in the study of subdivision surfaces. We consider a set of independent and preferably real eigenvectors of the subdivision matrix, aiming at deriving the local analytic properties of the limit surface. On the other hand, the focus of this paper is on the explicit decomposition of the initial shape into simple geometric components corresponding to the eigenvectors of the subdivision matrix. As we will discuss in section 10.2, such decompositions can reveal interesting geometric properties of the limit shape which are difficult to infer from its analytic properties.

One additional factor making the geometric decompositions of the shapes interesting, is that they are not unique. First, the eigenvectors of the matrix usually have high geometric multiplicity. Then, each eigenvector may be represented by more than one component of the sum in Eq. 10.1. Indeed, as we will see in section 10.5, the use of several copies of a geometric eigencomponent, placed in different positions in the 3d space, may lead to more intuitive decompositions of the initial shape.

10.1.2 Previous Work

The transformation of a polygon by joining the middles of adjacent edges to create a new polygon is simple geometric problem where circulant matrices arise. It was studied as early as 1878 by Darboux in [1]. Several generalizations of this problem have been studied and the connection between such transformations and circulant matrices is now well-understood [2, 3, 4, 5, 6]. Applications of this theory to the study of subdivision surfaces have been proposed in [7, 8].

Regarding the G-circulant matrices, their spectral properties are usually studied with the use of group *characters*. In this paper we keep the standard terminology and notation of [9], even though we deal with finite abelian groups only, and thus utilize a very small portion of the theory of group characters. Advanced results on the relation between G-circulant matrices and graphs can be found in [10], while the block-diagonalization of G-circulant matrices, where G is non-abelian, is studied in [11].

10.1.3 Overview

As motivation for the study of higher dimensional objects, in section 10.2 we discuss decompositions of polygons. In section 10.3 we briefly review the standard terminology from character theory and the basic theorems that allow us to compute the eigenvalues and eigenvectors of G-circulant matrices, where G is an abelian group. In section 10.4 we study decompositions of n-dimensional cubes by eigenvectors of subdivision matrices corresponding to multivariate quadratic splines. Finally, in section 10.5 we study the decompositions of prisms by the eigenvectors of a subdivision matrix corresponding to the tensor extension of the Doo-Sabin surface subdivision scheme.

10.2 Circulant matrices and polygonal decompositions

A matrix is called circulant if each row is the previous row cycled forward one step

$$\begin{pmatrix} c_0 & c_1 & c_2 & \cdots & c_{n-2} & c_{n-1} \\ c_{n-1} & c_0 & c_1 & \cdots & c_{n-3} & c_{n-2} \\ & & \cdots & & & \\ c_2 & c_3 & c_4 & \cdots & c_0 & c_1 \\ c_1 & c_2 & c_3 & \cdots & c_{n-1} & c_0 \end{pmatrix} \qquad (10.3)$$

The columns of the Discrete Fourier Transform matrix

$$F_n = \begin{pmatrix} 1 & 1 & 1 & \cdots & 1 \\ 1 & \omega & \omega^2 & \cdots & \omega^{n-1} \\ 1 & \omega^2 & \omega^4 & \cdots & \omega^{2(n-1)} \\ \vdots & \vdots & \vdots & & \vdots \\ 1 & \omega^{n-1} & \omega^{2(n-1)} & \cdots & \omega \end{pmatrix} \qquad (10.4)$$

where $\omega = e^{\frac{2\pi i}{n}}$, form a set of independent eigenvectors for every circulant $n \times n$ matrix. That means that the eigenvectors of a circulant matrix depend only on its dimension. We also notice that with one or two exceptions, depending on the parity of n, the eigenvectors come in conjugate pairs.

We geometrically interpret these eigenvectors as planar polygons, under the convention that by polygon we mean any n-tuple of points

$$(p_0, p_1, \ldots, p_{n-1}) \qquad (10.5)$$

thus, allowing multiple vertices and self-intersections. We will refer to the polygons corresponding to the eigenvectors of a matrix as *eigenpolygons*. The vertices of an eigenpolygon corresponding to a real eigenvector are collinear, while the vertices of an eigenpolygon corresponding to a complex eigenvector are coplanar.

Fig. 10.1 shows a set of eigenpolygons corresponding to an eigenbasis for the circulant matrices of dimension five and six. Notice that the choice of the eigenbasis is not unique, not even if we define each eigenvector up to a scalar. Indeed, especially in the circulant matrices appearing in subdivision applications, multiple eigenvalues are very common, and the corresponding eigenspaces have dimension higher than one. Here we chose the eigenbasis given by the columns of the Discrete Fourier Transform matrix shown in Eq. 10.4.

Every polygon embedded in $\mathbf{R}^2$ can be written as a unique linear combination of these eigenpolygons, because the corresponding complex vectors are linearly independent and form a basis of the space of n-dimensional complex vectors. The situation is more complicated with polygons embedded in $\mathbf{R}^3$. Such polygons are not necessarily planar, that is, not all their vertices are necessarily coplanar. However, in [2] it was shown that every polygon embedded in $\mathbf{R}^3$ can be written as a linear combination of planar eigenpolygons embedded in $\mathbf{R}^3$, with the additional property that any two conjugate eigenpolygons are coplanar. By writing such a decomposition, which is not necessarily unique, in the form of Eq. 10.1, we get

$$z_0 C_0 + z_1 C_1 + \cdots + z_{n-1} C_{n-1} \qquad (10.6)$$

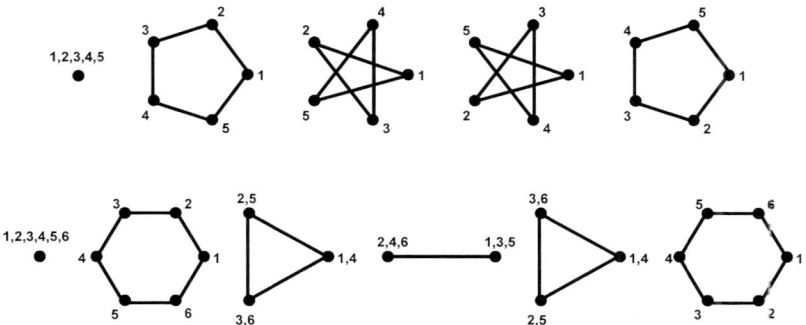

Fig. 10.1. Top: The eigenpentagons. **Bottom:** The eigenhexagons

where C_i are the planar polygons given by the columns of the Fourier matrix F_n, embedded in $\mathbf{R}^3$ and z_i are complex numbers. The complex multiplication $z_i C_i$ is understood in the plane of C_i, while the plus signs denote point addition in the 3d space.

In a typical subdivision scheme, for example, Loop [12], Butterfly [13] or Doo-Sabin [14], polygons corresponding faces or to 1-ring neighborhoods of vertices evolve to the next subdivision step by a multiplication by a circulant matrix. Fig. 10.4 shows an example of one step of Doo-Sabin subdivision. Recall that the Doo-Sabin subdivision scheme refines a polygonal mesh by inserting k new vertices for each old face of order k and connecting them with faces as shown in Fig. 10.2. We can see that there is a correspondence between the new faces and the old edges, vertices and faces. The positions of the new vertices are linear combinations of the vertices of the corresponding old face, with coefficients that only depend on the order of that face. For example, from the quadrilateral $P_0 P_1 P_2 P_3$ shown in Fig. 10.2 (right) we compute a new quadrilateral $P_0' P_1' P_2' P_3'$ with the points P_0, P_1, P_2, P_3 given as linear combinations of the P_0', P_1', P_2', P_3'. When we write this transformation in matrix form we get a circulant matrix whose first row gives the linear combination corresponding to P_0'.

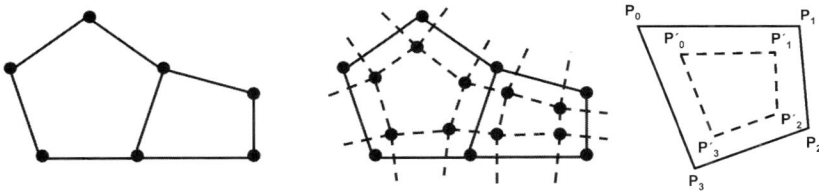

Fig. 10.2. Doo-Sabin subdivision.

In all of the subdivision schemes we mentioned above, the eigenvalue of the constant polygon C_0 is equal to one, the eigenvalues of the two regular affine polygons C_1, C_{n-1} are real, positive and equal, while all the other eigenvalues have smaller norms. That is,

$$1 = \lambda_0 > \lambda_1 = \lambda_{n-1} > |\lambda_i|, \quad i = 2, 3, \ldots, n - 2 \tag{10.7}$$

As the constant eigenpolygon corresponds to the eigenvalue with the largest norm, it follows from Eq. 10.2 that in the limit the polygon converges to a single point, which is its barycenter. The eigenpolygons with the next two largest eigenvalues λ_1, λ_{n-1} will determine the limit shape of the polygon. To define this limit shape explicitly, we first assume that the barycenter of the polygon is the origin, eliminating thus the first component of the sum in Eq. 10.2 and then we scale the polygon by a factor of $1/\lambda_1^m$ to counter the shrinkage effect. Eq. 10.2 becomes

$$(\lambda_1/\lambda_1)^m z_1 C_1 + (\lambda_2/\lambda_1)^m z_2 C_2 + \cdots + (\lambda_{n-1}/\lambda_1)^m z_{n-1} C_{n-1} \tag{10.8}$$

giving

$$A = z_1 C_1 + z_{n-1} C_{n-1} \tag{10.9}$$

as $m \rightarrow \infty$. Thus, the limit shape is the sum of two coplanar regular polygons with opposite orientations. In particular, A is the affine image of a regular polygon and can be inscribed in an ellipse with semi-axes $|z_1| + |z_{n-1}|$ and $||z_1| - |z_{n-1}||$, see [5]. If the eigenvalues λ_1, λ_{n-1} are complex, as it is the case with the *simplest scheme* proposed in [15], then we can still study the limit of Eq. 10.8, but in this case it might not exist. In the literature, the simplest schemes is analyzed by combining two steps to obtain a binary refinement step with real eigenvalues.

The geometric interpretation of Eq. 10.9 is simple and insightful, justifying, in our opinion, the choice of complex rather than real eigenvectors. For example, we can immediately see that A degenerates into a line when one of the two semi-axes of the ellipse has zero length, that is, when the two regular affine components have equal norms. In this case the subdivision surface will have a singularity at the point of convergence of the polygon. More interestingly, we can detect a second type of singularity by noticing that A has the same orientation as the component with the largest norm, cf. Fig. 10.3. In fact, this second type of singularity has higher dimension than the first in the space of planar polygons, even though such badly shaped non-convex polygons rarely appear in practical applications.

Notice that the comparison between the orientation of A and its two components is possible because they are all coplanar polygons. In fact, they are all on the tangent plane at the point of convergence of the initial polygon. In the case of planar meshes, we can also compare between the orientation of the mesh, its faces and their eigencomponents. In [7] it was shown an example of a polygon, which was part of a planar, consistently oriented mesh without self-intersections, and the larger affine regular component had orientation opposite to the mesh. That means that the limit shape of the polygon also had orientation opposite to the mesh, inverting the direction of the normal at that point of the plane. Of course, these types of singularities can also be studied with the use of a real eigenbasis. However, we believe that it is much more difficult to detect and classify them without a geometrically intuitive decomposition of the initial input.

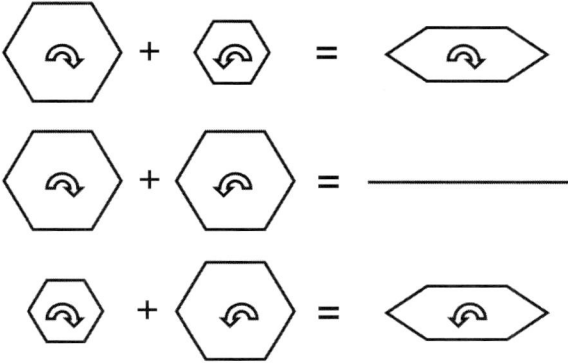

Fig. 10.3. Polygons can be added as vectors. The orientation of each polygon is shown by the arrow inside it. Notice that if the sum of two regular coplanar polygons does not degenerate, then it has the same orientation as the component with the larger norm.

The polygonal decomposition we studied above corresponds to a small part of the subdivision mesh, that is, a face or the 1-ring neighborhood of a vertex. To obtain results related to higher order properties of the limit surface, we have to study decompositions of larger pieces of the subdivision mesh. The main technical difficulty in constructing such decompositions is that the corresponding matrices are circulant block rather than circulant. In [8], decompositions of larger areas of a subdivision mesh are studied in the context of the Catmull-Clark scheme.

10.3 *G*-circulant matrices

A matrix is called G-circulant if each row is obtained from the first row by a permutation by an element of a group G, and the rows and columns of the matrix are indexed by the elements of G. If the group $G = \mathbf{Z}_n$ is cyclic, then we obtain the usual circulant matrices.

First we give some background theorems which allow the computation of the eigenvalues and eigenvectors of a G-circulant matrix, where G is a finite abelian group. Notice that the groups $\mathbf{Z}_2^n$ and $\mathbf{Z}_2 \times \mathbf{Z}_n$ we are interested in, are finite abelian.

Proposition 1. *Let G be a finite abelian group. Let the first row of the G-circulant matrix S be $(a_g)_{g \in G}$. Then, the eigenvalues of S are given by*

$$\lambda_\chi = \sum_{g \in G} \chi(g) a_g \tag{10.10}$$

and the eigenvectors (which depend only on G) are given by

$$\mathbf{v}_\chi = (\chi(g))_{g \in G} \tag{10.11}$$

where χ is an (irreducible) character of G.

The characters of a group are complex valued functions

$$\chi : G \to \mathbf{C} \tag{10.12}$$

If G is abelian, then there is an 1-1 correspondence between the characters and the elements of G. From Proposition 1 we can easily compute the eigenvalues and eigenvectors of a G-circulant matrix, as long as we know the characters of G. Given that every finite abelian group is the direct product of cyclic groups, the following two propositions giving the characters of a cyclic group and the characters of a direct product group suffice for our purposes.

Proposition 2. *The characters of the cyclic group $\mathbf{Z}_n$ are given by*

$$\chi_j : \chi_j(k) = \omega^{jk}, \quad \forall k \in \mathbf{Z}_n \tag{10.13}$$

where j is an element of $\mathbf{Z}_n$ and $\omega = e^{\frac{2\pi i}{n}}$

Proposition 3. *If a group G has a direct product structure, then its characters are tensor products of the characters of its components.*

10.3.1 G-circulant matrices and geometric transformations

G-circulant matrices can be used to study linear transformations of a mesh. We first index the rows and the columns of the matrix by the vertices of the mesh. Similarly to the case of circulant matrices and polygons in section 10.2, if the position of the new vertex P_0' is computed by the first row of the matrix, and if P_0' is mapped on P_i' by a symmetry g of the mesh, we expect that the row giving P_i' is the permutation of the first row by g.

In a general setting, we assume that a group of symmetries G acts on the vertices of a mesh in a way that every point is mapped on any other point by a unique element of G. In other words we have a free transitive action. This is the case with the n-dimensional cube and the prism, and the groups $\mathbf{Z}_2^n$ and $\mathbf{Z}_2 \times \mathbf{Z}_n$, respectively. Notice that the choice of the group G is not necessarily unique. For example, for the quadrilateral we can either use the $\mathbf{Z}_2^2$ or the $\mathbf{Z}_4$, while for the hexahedron we can use the $\mathbf{Z}_2^3$ or the $\mathbf{Z}_2 \times \mathbf{Z}_4$. With either choice, the transformations and the matrices are the same up to a labeling, but Proposition 1 yields different sets of eigenvectors and thus, we obtain different geometric interpretations.

10.4 Cube decompositions

To apply the above setup to the n-dimensional cube, we first need a correspondence between its vertices and the elements of $\mathbf{Z}_2^n$. This can be easily done by considering the unit cube. Indeed, the coordinates of its vertices are n-tuples with entries 0's and 1's, giving the corresponding elements of $\mathbf{Z}_2^n$, cf. Fig. 10.4. For referencing a specific component of the n-tuple of g we use the characteristic functions δ_i, $i = 1, 2, \ldots, n$

$$\delta_i(g) = \text{the } i\text{th value of the } n\text{-tuple of } g \qquad (10.14)$$

We also define $\sigma(g)$ as the number of 1's in the n-tuple

$$\sigma(g) = \sum_{i=1}^{n} \delta_i(g) \qquad (10.15)$$

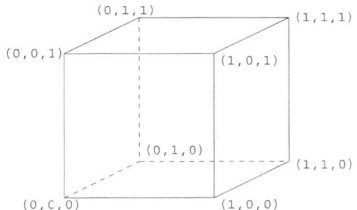

Fig. 10.4. Labeling the vertices of a cube with the elements of $\mathbf{Z}_2^3$.

Table 10.1 shows the eigenvectors of the $\mathbf{Z}_2^n$-circulant matrices for $n = 2, 3$, normalized by a factor of $(1/2^n)$. A brief description of the relevant character computations can be found in the Appendix. Notice that Proposition 1 gives an indexing of the eigenvectors by the characters of G, which in turn gives an indexing by the elements of G. Also notice that the entries of all the eigenvectors are either 1 or -1. This is a property that holds for arbitrary n. In fact, the eigenvectors of the $\mathbf{Z}_2^n$-circulant matrices are given by the columns of the familiar Walsh-Hadamard matrix H_n, see [10].

Because the above eigenvectors are linearly independent and real they can immediately be used for a geometric decomposition of an n-dimensional cube i.e., for the decomposition of an 2^n-tuple of n-dimensional points. To find the decomposition we multiply the 2^n-tuple (P_g), $g \in G$ by the inverse of the eigenvector's matrix. If the transformed 2^n-tuple is (P'_g), $g \in G$, then the decomposition of the initial cube is

$$\sum_{g \in G} P'_g \mathbf{v}_g \qquad (10.16)$$

where $\mathbf{v}_g$ is the row of the matrix corresponding to g.

00	1	1	1	1
10	1	-1	1	-1
01	1	1	-1	-1
11	1	-1	-1	1

/4

000	1	1	1	1	1	1	1	1
100	1	-1	1	1	-1	-1	1	-1
010	1	1	-1	1	-1	1	-1	-1
001	1	1	1	-1	1	-1	-1	-1
110	1	-1	-1	1	1	-1	-1	1
101	1	-1	1	-1	-1	1	-1	1
011	1	1	-1	-1	-1	-1	1	1
111	1	-1	-1	-1	1	1	1	-1

/8

Table 10.1. The eigenvectors of $\mathbf{Z}_2^2$ and $\mathbf{Z}_2^3$ are the rows of the tables.

Similarly to the polygonal case, the point $P'_{(0,\ldots,0)}$ is the barycenter of the cube and we may assume it to be the origin. Then, there are n eigenvectors with $\sigma(g) = 1$, e.g. 10 and 01 for $n = 2$ and 100, 010 and 001 for $n = 3$. The components of these eigenvectors are equal to $1/2^n$ at all vertices of one face, and $-1/2^n$ at all vertices of the opposite face. Thus, a point P'_g with $\sigma(g) = 1$ is at the barycenter of a face, which means that the barycenter of the opposite face is at $-P'_g$. Generally, there are $\binom{n}{k}$ elements of G with $\sigma(g) = k$.

Below we study in more detail the cases $n = 2, 3$.

The quadrilateral (n=2): Using the geometrically intuitive cyclic ordering of the vertices instead of ordering them by the value of σ as above, let the initial quadrilateral be $(P'_{00}, P'_{10}, P'_{11}, P'_{01})$. The component $(P'_{00}, P'_{00}, P'_{00}, P'_{00})$ is its barycenter and we may assume it is the origin. The next components, $(P'_{10}, P'_{10}, -P'_{10}, -P'_{10})$ and $(P'_{01}, -P'_{01}, -P'_{01}, P'_{01})$ join the middles of opposite edges of the quad. If the fourth component $(P'_{11}, -P'_{11}, P'_{11}, -P'_{11})$ is zero, then the quad is a parallelogram, with $(P'_{10}, -P'_{10})$ and $(-P'_{01}, P'_{01})$ giving its two directions. The magnitude of the fourth component can be thought as a measure of how far is the quad from being a parallelogram. Notice that $(P'_{11}, -P'_{11})$ joins the middles of the diagonals of the quad, while a quad is a parallelogram if and only if its diagonals bisect. A quad is planar if and only if P'_{10}, P'_{01} and P'_{11} are linearly dependent.

It is interesting to compare the decomposition of the quad given by $G = \mathbf{Z}_2^2$ with the one obtained with $G = \mathbf{Z}_4$, cf. Fig. 10.5. The difference in the two decompositions is the use of the eigenvectors $(1, i, -1, -i)$ and $(1, -i, -1, i)$ instead of $(1, 1, -1, -1)$ and $(1, -1, -1, 1)$. The former, represent two squares with opposite orientation and a linear combination of them is the affine image of a square, i.e., a parallelogram [2].

The hexahedron (n=3): Similarly to the case $n = 2$, P_{000} is the barycenter of the hexahedron while the three components with $\sigma(g) = 1$ give the three directions of a parallelepiped. The hexahedron is a parallelepiped if and only if the four other components are zero. The geometric interpretation of the first three of these four conditions is recursively related to the two dimensional case. Indeed, a hexahedron has twelve edges, which can be separated into three subsets of four edges with the same

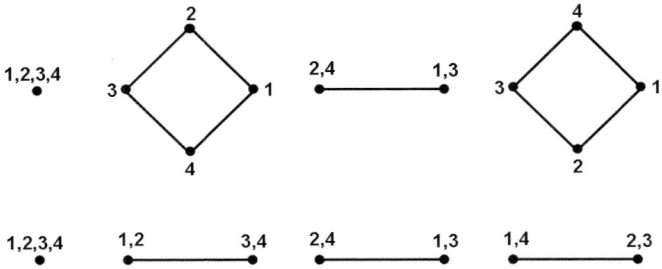

Fig. 10.5. The eigencomponents of a $\mathbf{Z}_4$-circulant matrix (top) and a $\mathbf{Z}_2^2$-circulant matrix bottom.

direction, each one corresponding to an eigencomponent with $\sigma(g) = 1$, cf. Fig. 10.6 (left). If we want the corresponding eigencomponent to be zero, the middles of the segments joining the middles of opposite edges should be the same. That means that for each subset, the middles of the four edges should form a parallelogram.

The fourth condition says that the barycenter of the points with $\sigma(g) = 0, 2$ should be the same with the barycenter of the points with $\sigma(g) = 1, 3$, cf. Fig. 10.6 (right). This is similar to the $n = 2$ case, where a quadrilateral is a parallelogram if a only if the points with $\sigma(g) = 0, 2$ and the points with $\sigma(g) = 1$ have the same barycenter.

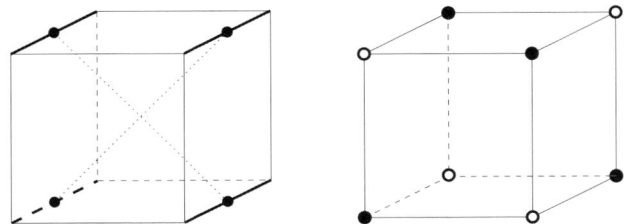

Fig. 10.6. Left: Four edges with the same direction are shown in bold. **Right:** The vertices with $\sigma(g) = 0, 2$ (filled circles) and $\sigma(g) = 1, 3$ (empty circles).

From the above discussion it is intuitively clear that the existence of large components with $\sigma(g) = 2, 3$ leads to hexahedra with convoluted shapes. Fig. 10.7 shows the effect of adding a single non-zero component with $\sigma(g) = 2, 3$ to the unit cube.

10.4.1 Application: the multivariate quadratic spline

Up to now we studied general $\mathbf{Z}_2^n$-circulant matrices without any reference to their elements. In this section we study the $\mathbf{Z}_2^n$-circulant matrix corresponding to the evolution of one cell of an n-dimensional grid under the quadratic B-spline subdivision

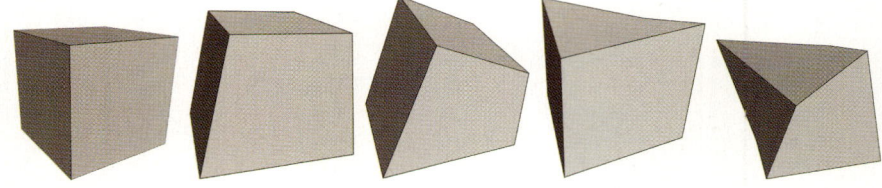

Fig. 10.7. From left to right: (a) The unit cube. (b) $P_{110} = (0.2,\ 0.0,\ 0.0)$. (c) $P_{110} =$ $(0.2,\ 0.2,\ 0.0)$. (d) $P_{110} = (0.0,\ 0.0,\ 0.2)$. (e) $P_{111} = (0.2,\ 0.2,\ 0.2)$.

algorithm. The row of the matrix giving the new position of the vertex corresponding to g is

$$a_g = \frac{3^{n-\sigma(g)}}{4^n} \tag{10.17}$$

For example, if $n = 2$ we get the matrix

$$\begin{pmatrix} 9\ 3\ 3\ 1 \\ 3\ 9\ 1\ 3 \\ 3\ 1\ 9\ 3 \\ 1\ 3\ 3\ 9 \end{pmatrix} /16 \tag{10.18}$$

We have

Proposition 4. *The eigenvalue corresponding to the eigenvector* $\mathbf{v}_g$ *of the subdivision matrix of the n-dimensional quadratic spline is* $\frac{1}{2^{\sigma(g)}}$.

For a sketch of the proof, we notice by Eq. 10.10, 10.17 the eigenvalue corresponding to the character χ_g is

$$\lambda_{\chi_g} = \sum_{h \in G} \chi_g(h) \frac{3^{n-\sigma(h)}}{4^n} \tag{10.19}$$

giving,

$$\lambda_{\chi_g} = \frac{(3+1)^{n-\sigma(g)}(3-1)^{\sigma(g)}}{4^n} \tag{10.20}$$

To see this, we expand the product $(3+1)^{n-\sigma(g)}(3-1)^{\sigma(g)}$ and rearrange the factors so that the terms $(3\text{-}1)$ are placed at the positions where $\delta(g) = 0$. Finally, from Eq. 10.20 we get

$$\lambda_{\chi_g} = \frac{1}{2^{\sigma(g)}} \tag{10.21}$$

□

In the limit, the cell converges to a single point, which is its barycenter. Assuming that the barycenter is the origin, the limit shape is given by the eigencomponents of the next eigenvalues, that is by the n components with eigenvalue 1/2. After scaling the cell to counter the shrinkage effect we get

Proposition 5. *Under multivariate quadratic B-spline subdivision the limit shape of a cell is the sum of the eigencomponents with $\sigma(g) = 1$. In particular, it is a parallelogram for $n = 2$ and a parallelepiped for $n = 3$.*

Similarly to the polygonal case, we can use the decomposition to find when singularities appear at the point of convergence of the initial cell. For example. in the case $n = 3$, if two opposite faces of the initial hexahedron have the same barycenter, then the parallelepiped given by the three eigenvectors with eigenvalue $1/2$ will collapse to a parallelogram. A different type of singularity appears when the orientation of one of the limit shape parallelepipeds is not consistent with the rest of the grid. However, it should be noted that even though we can study singularities at the barycenters of the cells, the method can not be used to deduce any analytic prcperties of the limit volume, because we study the evolution of one cell in isolation.

10.5 Prism decomposition

Next we study the decomposition of a prism by the eigenvectors of the $\mathbf{Z}_2 \times \mathbf{Z}_n$-circulant matrices. By Eq. 10.11 these eigenvectors are the rows of the matrix

$$
\begin{bmatrix}
1 & 1 & 1 & \cdots & 1 & 1 & 1 & 1 & \cdots & 1 \\
1 & \omega & \omega^2 & \cdots & \omega^{n-1} & 1 & \omega & \omega^2 & \cdots & \omega^{n-1} \\
1 & \omega^2 & \omega^4 & \cdots & \omega^{2n-2} & 1 & \omega^2 & \omega^4 & \cdots & \omega^{2n-2} \\
\cdots & \cdots & \cdots & \cdots & \cdots & & \cdots & & & \\
1 & \omega^{n-1} & \omega^{n-2} & \cdots & \omega & 1 & \omega^{n-1} & \omega^{n-2} & \cdots & \omega \\
1 & 1 & 1 & \cdots & 1 & -1 & -1 & -1 & \cdots & -1 \\
1 & \omega & \omega^2 & \cdots & \omega^{n-1} & -1 & -\omega & -\omega^2 & \cdots & -\omega^{n-1} \\
\cdots & \cdots & \cdots & \cdots & \cdots & & \cdots & & & \\
1 & \omega^{n-1} & \omega^{n-2} & \cdots & \omega & -1 & -\omega^{n-1} & -\omega^{n-2} & \cdots & -\omega
\end{bmatrix}
\qquad (10.22)
$$

Fig. 10.8 shows the eigenvectors corresponding to the rows 1, $n - 1$, $n + 1$ and $2n - 1$ (the enumeration of the rows starts from zero). Because the eigenvectors are complex, these eigenprisms are planar. For that reason it is not straightforward to find a formula similar to Eq. 10.16 where all the eigenvectors were real.

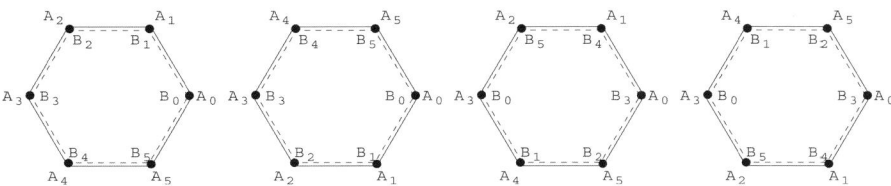

Fig. 10.8. The eigenprisms given by the rows 1, $n - 1$, $n + 1$, $2n - 1$ of the matrix Notice that the multiplication of a polygonal face by -1 corresponds to a rotation by π.

Instead, we build a construction based on the decompositions of non-planar polygons proposed in [2]. Recall that every non-planar n-gon can be written in the form shown in Eq. 10.6. These polygons generally lie on different planes but C_j and C_{n-j} are always coplanar.

Let A, B be the two n-gonal faces of the prism and let their decompositions be

$$A = a_0 C_0 + a_1 C_1 + \cdots + a_{n-1} C_{n-1}$$
$$B = b_0 C_0 + b_1 C_1 + \cdots + b_{n-1} C_{n-1} \tag{10.23}$$

Notice that the copies of C_i used in the decomposition of A and B generally lie on different planes of the 3-dimensional space. By concatenating the two polygonal decompositions we get a prism decomposition

$$[a_0 C_0, b_0 C_0] + [a_1 C_1, b_1 C_1] + \cdots + [a_{n-1} C_{n-1}, b_{n-1} C_{n-1}] \tag{10.24}$$

For simplicity we only deal with the components C_1, C_{n-1}, as all the other conjugate pairs, as well as the single components C_0 and $C_{n/2}$ (for n even) can be treated similarly.

Let E_A be the plane of the components $a_0 C_0, a_{k-1} C_{k-1}$ lie and let E_B be the plane of the components $b_0 C_0, b_{k-1} C_{k-1}$. Working first on the E_A plane we write the components $[a_1 C_1, 0]$ and $[a_{n-1} C_{n-1}, 0]$ as a linear combination of the four eigenprisms of Fig. 10.8, i.e. as

$$x_1 [C_1, C_1] + x_{n-1} [C_{n-1}, C_{n-1}] + x_{n+1} [C_1, -C_1] + x_{2n-1} [C_{n-1}, -C_{n-1}] \tag{10.25}$$

We get

$$x_1 C_1 + x_{n+1} C_1 = a_1 C_1$$
$$x_1 C_1 - x_{n+1} C_1 = 0 \cdot C_1$$
$$x_{n-1} C_{n-1} + x_{2n-1} C_{n-1} = a_{n-1} C_{n-1}$$
$$x_{n-1} C_{n-1} + x_{2n-1} C_{n-1} = 0 \cdot C_{n-1} \tag{10.26}$$

giving

$$x_1 = x_{n+1} = \frac{a_1}{2} \qquad x_{n-1} = x_{2n-1} = \frac{a_{n-1}}{2} \tag{10.27}$$

Similarly, working with the components $[0, b_1 C_1]$ and $[0, b_{n-1} C_{n-1}]$ we get four more eigenprisms, this time on the E_B plane, with

$$x_1 = x_{n+1} = \frac{b_1}{2} \qquad x_{n-1} = x_{2n-1} = \frac{b_{n-1}}{2} \tag{10.28}$$

We notice that the obtained decomposition is quite heavy as we use eight eigenprisms for just four polygonal components. However, the eigenvalue λ_1 corresponding to the $[C_1, C_1]$ and $[C_{n-1}, C_{n-1}]$ components is usually larger than the eigenvalue of the $[C_1, -C_1]$ and $[C_{n-1}, -C_{n-1}]$ components. Thus, the limit shape of the prism will be determined by fewer than eight components.

Indeed, this is the case with the eigenvalues corresponding to the tensor product of the Doo-Sabin subdivision rule. Under this subdivision scheme, the limit shape of the prism is determined by five components corresponding to the second largest eigenvalue λ_1. These are the components $[C_1, C_1]$ and $[C_{n-1}, C_{n-1}]$

These are the components $[C_1, C_1]$ and $[C_{n-1}, C_{n-1}]$ on the planes E_A and E_B, which are four in total, and the component corresponding to nth row of the matrix in Eq. 10.22. The eigenvalue λ_n of this component should also be equal to λ_1. Otherwise, the ratio between the height of the prism and the diameter of its base will tend to 0 or to ∞, depending on whether λ_n is smaller or larger than or λ_1.

Eq.(10.25) gives

$$[\frac{(a_1 + b_1)C_1 + (a_{n-1} + b_{n-1})C_{n-1}}{2}, \frac{(a_1 + b_1)C_1 + (a_{n-1} + b_{n-1})C_{n-1}}{2}]$$

$$(10.29)$$

where the use of the letter a or b in the coefficient also indicates the plane of the component. We notice that A and B have the same limit shape. Moreover, the limit shape of A, B is planar and thus, the limit shape of the prism is regular. Fig. 10.9 shows the evolution of a pentagonal prism under this subdivision scheme. An outline for the explicit computations of A and B is shown in the Appendix.

10.6 Conclusion - Future Work

We studied decompositions of cubes and prisms by the eigenvectors of G-circulant matrices. We concentrated on the geometric interpretations of these decompositions and we studied the evolution of single cells under linear transformations. As an application we obtained information about the singularities in quadratic n-dimensional splines.

In the future we plan to extend our work to the study of evolutions of larger configurations, instead of the single cells we are currently dealing with. Such a generalization will allow the study of singularities in higher degree splines and general volume subdivision grids.

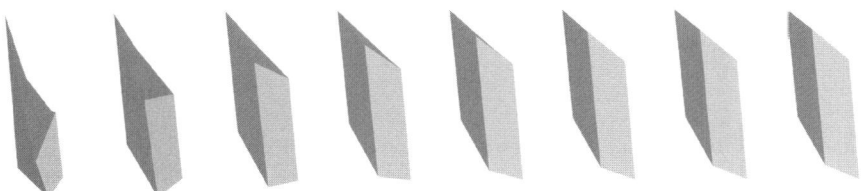

Fig. 10.9. The evolution of a pentagonal prism.

10.7 Appendix

A. The eigenvectors of $\mathbf{Z}_2^n$
By Proposition 2 the two characters of $\mathbf{Z}_2$ are

$$\chi_0 : \chi_0(0) = 1, \chi_0(1) = 1 \tag{10.30}$$

and

$$\chi_1 : \chi_1(0) = 1, \chi_1(1) = -1 \tag{10.31}$$

By Proposition 3 the character χ_h corresponding to the element h of $\mathbf{Z}_2^n$ is given by

$$\chi_h(g) : \prod_{i=1}^{n} e_{ihg} \tag{10.32}$$

where $e_{ihg} = -1$ iff $\delta_i(h) = \delta_i(g) = 1$ and $e_{ihg} = 1$ otherwise.
Finally, we can use Proposition 1 to compute the eigenvectors of the $\mathbf{Z}_2^n$-circulant matrices.

B. Exact computation of the limit shape of the prism
To compute the exact limit shape of A, B we write the polygons $\frac{a_1}{2} C_1 + \frac{a_{n-1}}{2} C_{n-1}$ and $\frac{b_1}{2} C_1 + \frac{b_{n-1}}{2} C_{n-1}$ in parametric form. We notice that they are both planar polygons inscribed on ellipses, thus their vertices lie on the curves

$$\mathbf{c_a} + r_{au} \cos \theta \mathbf{u_a} + r_{av} \sin \theta \mathbf{v_a} \tag{10.33}$$

and

$$\mathbf{c_b} + r_{bu} \cos \theta \mathbf{u_b} + r_{bv} \sin \theta \mathbf{v_b} \tag{10.34}$$

respectively, where $\mathbf{c_a}, \mathbf{c_b}$ are the centers of the ellipses, $\mathbf{u_a}, \mathbf{v_a}$ and $\mathbf{u_b}, \mathbf{v_b}$ are orthonormal vectors on E_A, E_B and $\theta = \frac{2\pi j}{k}$, $j = 0, 1, \ldots, k - 1$.
Then Eq.(10.29) gives the limit shape as

$$(\mathbf{c_a} + \mathbf{c_b}) + \cos \theta (r_{au} \mathbf{u_a} + r_{bu} \mathbf{u_b}) + \sin \theta (r_{av} \mathbf{v_a} + r_{bv} \mathbf{v_b}) \tag{10.35}$$

which is again the equation of an ellipse.

References

1. Darboux, G.: Sur un problème de géométrie élémentaire. Darboux Bull. (1878)
2. Berlekamp, E., Gilbert, E., Sinden, F.: A polygon problem. Am. Math. Mon. **72** (1965) 233–241
3. Bachmann, F., Schmidt, E.: n-Gons. University of Toronto Press (1975)
4. Davis, P.J.: Circulant matrices. Wiley-Interscience. (1979)
5. Fisher, J., Ruoff, D., Shilleto, J.: Perpendicular polygons. Am. Math. Mon. **92** (1985) 23–37

6. Bruckstein, A., Sapiro, G., Shaked, D.: Evolutions of planar polygons. IJPRAI **9** (1995) 991–1014
7. Ivrissimtzis, I., Seidel, H.P.: Evolutions of polygons in the study of subdivision surfaces. Computing **72**(1-2) (2004) 93–104
8. Ivrissimtzis, I., Zayer, R., Seidel, H.P.: Polygonal decomposition of the 1-ring neighborhood of the Catmull-Clark scheme. In: Proceedings of the Shape Modeling International, IEEE (2004) 101–109
9. Isaacs, I.: Character theory of finite groups. New York, Dover Publications (1994)
10. Adamczak, W., Andrew, K., Hernberg, P., Tamon, C.: A note on graphs resistant to quantum uniform mixing. preprint (arXiv:quant-ph/0308073 v1) (2003)
11. Diaconis, P.: Patterned matrices. In: Matrix theory and applications, Proc. Symp. Appl. Math. 40, 37-58. (1990)
12. Loop, C.T.: Smooth subdivision surfaces based on triangles. Master's thesis, University of Utah, Department of Mathematics (1987)
13. Dyn, N., Levine, D., Gregory, J.A.: A butterfly subdivision scheme for surface interpolation with tension control. ACM ToG **9**(2) (1990) 160–169
14. Doo, D., Sabin, M.: Behaviour of recursive division surfaces near extraordinary points. Computer-Aided Design **10** (1978) 356–360
15. Peters, J., Reif, U.: The simplest subdivision scheme for smoothing polyhedra. ACM ToG **16**(4) (1997) 420–431

Subdivision Methods for the Topology of 2d and 3d Implicit Curves

Chen Liang, Bernard Mourrain, and Jean-Pascal Pavone

GALAAD, INRIA
BP 93, 06902 Sophia Antipolis, France
{mourrain, jppavone}@sophia.inria.fr

Summary. In this paper, we describe a subdivision method for handling algebraic implicit curves in 2d and 3d. We use the representation of polynomials in the Bernstein basis associated with a given box, to check if the topology of the curve is determined inside this box, from its points on the border of the box. Subdivision solvers are used for computing these points on the faces of the box, and segments joining these points are deduced to get a graph isotopic to the curve. Using envelop of polynomials, we show how this method allow to handle efficiently and accurately implicit curves with large coefficients. We report on implementation aspects and experimentations on 2d curves such as ridge curves or self intersection curves of parameterized surfaces, and on silhouette curves of implicit surfaces, showing the interesting practical behavior of this approach.

11.1 Introduction

In this paper, we address the problem of computing the topology of 3D curves resulting from the intersection of two algebraic surfaces. Algebraic curves and surfaces are compact representations of shapes, which can be complex and have numerous advantages over parametric ones, such as easy determination of inside/outside of the surface. This is particularly useful when we have to apply logical operations (union, subtraction, etc.) between two solid objects, defined implicitly. In such problems, computing the intersection of two surfaces is a critical operation, which has to be performed efficiently and accurately. Implicit curves and surfaces have also disadvantages such as difficulty in performing graphical display, but the method that we propose in this paper is step towards handling such problems, since it allows fast display of this implicitly 2d and 3d curves. On the other hand, dealing with parameterized surfaces naturally leads to the computation of implicit curves. Let us mention in particular, the computation of the intersection curve of two surfaces, self-intersection curves, plane sections and ridge curves (which are defined implicitly on these parameterised surfaces, though they are usually approximated by parameterised curves). Such problems reduce to the analysis of a curve defined by $n - 1$ polynomial equations, in a space of dimension n (here $n = 2, 3, 4$).

One major obstacle for adopting implicit representations instead of parametric representations concerns the piecewise linear approximation of such curves or surfaces for visualization purposes. A brute force approach would be an exhaustive evaluation for approximating the zero level set, which is obviously very inefficient. A typical alternative scenario is to adopt a divide-and-conquer approach. Larger undetermined domains are broken down to smaller predictable domains in which the topological feature and eventually, the curve/surface itself can be inferred efficiently. An objective of this paper is to describe an efficient method, which allows us to capture the topology of an implicit curve, when this curve is smooth[1], but also to localize the singular points if they exist.

The problem of computing the topology of curves has been approach in different ways. A first family of methods is based on a sweeping approach. For 2D planar algebraic curves, such approach has been studied in [7] and [10]. It was later extended by Gatellier et al. in [8] to the 3D spatial curves resulting from the intersection of two algebraic surfaces. See also [2]. These methods use a conceptual sweeping line/plane perpendicular to some projection axis, and detect the critical topological events, such as tangents to the sweeping planes and singularities. The final output of these methods are a graph of connected vertices complying to the topology of the original curves. A notable problem of aforementioned approaches is that they relies of the computation of sub-resultant sequences, which can be a bottleneck in many examples with large degree and large coefficients (see Section 11.4.1).

Another family of methods are the subdivision based techniques, which uses a simple criterion to remove domains which do not contain the roots. A crucial problem involved here is how to efficiently and reliably deduce the root information in a given interval (or a bounding box). In these methods, instead of using monomial representation, we represent the equations using Bernstein basis [6]. Among early attempts, Sederberg [17] converted an algebraic curve in to piecewise triangular Bernstein basis. See also [13] combining symbolic and numeric techniques to compute the topology of 2D curves. The approach of [11] for computing the curves of intersection of two parameterised surfaces is also combining subdivision techniques with regularity criterion, exploiting the properties of the intersection curve in the 2D parameter domains.

The first problem of computing roots of univariate polynomials has been analyzed for instance in [15], where root information tests are by based on *Descartes' Law of Sign* and its variant in the Bernstein basis. This approach has been extended to the approximation of isolated roots of multivariate systems. In [18], the author used tensor product version of Bernstein basis and integrated domain reduction techniques to speed up the convergence and reduce the number of subdivisions. In [4], the emphasis is put on the subdivision process, and stopping criterion based on the normal cone to the surface patch. In [14], this approach has been improved by introducing pre-conditioning and univariate-solver steps. The complexity of the method is also analyzed in terms of intrinsic differential invariants.

---

[1] The tangent vector space exists at every points

The application of subdivision methods for handling higher dimensional objects is not so well developed. In [12] a method which subdivides up to some precision level, and applies dual marching cube approach to connect points on the curve or to mesh a surface is described. The variety is covered by boxes of a given size, and the connectivity of these cells is used to deduce the piecewise linear approximation. In [1], a subdivision approach exploiting the sign variation of the coefficients in the Bernstein basis in order to certify the topology of the surface in a cell, is used for the purpose of polygonalizing an implicit algebraic surface.

The work of this paper is in the spirit of this former approach. We apply a subdivision approach also exploiting the properties of the Bernstein polynomial representation. We describe a simple regularity test extending the criterion cf [1] to curves, which allows us to detect easily when the topology of the curve in a cell is uniquely determined from its intersection with the border of the cell. This provides an efficient test for stopping earlier the subdivision process and branching to path following methods if we are interested in a good geometrical approximation of the curve.

We address the same question as in [8], but with this new methods, we are able to solve the following problems already identified in this paper:

- To achieve higher numerical stability by operating on Bernstein basis instead of monomial basis;
- Through subdivision on three principle directions, i.e. x, y, z (or x, y, z), to isolate the domain containing the singularities from those containing regular curve segments. This divide-and-conquer approach, in principle, should simplify the graph building algorithm adopted in [8] where the whole domain has to be considered.

However, for the treatment of singular points, we have to introduced a threshold ϵ to stop the subdivision. Contrarily to [8], we do not certify the topology at singular points, but computed boxes of size ϵ, containing these singularities.

On the contrary, we show that our approach is able to handle implicit curves with large equation (of total degree about 80 with coefficients of bit-size 200), which resultant-based techniques are not able to treat.

This paper is organized as follows: in Section 11.2, we will review some of the relevant concepts and theorem required by our proposed algorithm; Section 11.3 is devoted to outlining our proposed algorithms and the details about how the essential steps in our algorithm are handled. We will show the experiment results in Section 11.4 and conclude in Section 11.5 with the problems and possible improvements over the currently proposed algorithm.

11.2 Fundamental ingredients

This section introduces the theoretical background of Bernstein polynomial representation and how it is related to the problem we want to solve. For a domain $D \subset \mathbb{R}^n$, we denote by $\overset{\circ}{D}$ its interior, by $\overline{D}$ its closure. For a box $D = [a_0, b_0] \times [a_1, b_1] \times$

$[a_2, b_2] \subset \mathbb{R}^3$, its x-face (resp. y-face, z-face) are its faces normal to the direction x (resp. y, z).

11.2.1 Univariate Bernstein basis

Given an arbitrary univariate polynomial function $f(x) \in \mathbb{K}$, we can convert it into the representation of Bernstein basis of degree d, which is defined by:

$$f(x) = \sum_i b_i B_i^d(x), \text{ and} \tag{11.1}$$

$$B_i^d(x) = \binom{d}{i} x^i (1-x)^{d-i} \tag{11.2}$$

where b_i is usually referred as controlling coefficients. Such conversion is done through a basis conversion [6]. The above formula can be generalized to an arbitrary interval $[a, b]$ by a variable substitution $x' = (b-a)x + a$. We denote by $B_d^i(x; a, b)\binom{d}{i}(x-a)^i(b-x)^{d-i}(b-a)^{-d}$ the corresponding Bernstein basis on $[a, b]$.

There are several useful properties regarding Bernstein basis given as follows:

- *Convex-Hull Properties*: Since $\sum_i B_d^i(x; a, b) \equiv 1$ and $\forall x \in [a, b]$, $B_d^i(x; a, b) \geq 0$ where $i = 0, ..., d$, the graph of $f(x) = 0$, which is given by $(x, f(x))$, should always lie within the convex-hull defined by the control coefficients [5].
- *Subdivision* (de Casteljau): Given $t_0 \in [0, 1]$, $f(x)$ can be represented piecewisely by:

$$f(x) = \sum_{i=0}^{d} b_0^{(i)} B_d^i(x; a, c) = \sum_{i=0}^{d} b_i^{(d-i)} B_d^i(x; c, b), \text{ where} \tag{11.3}$$

$$b_i^{(k)} = (1 - t_0) b_i^{(k-1)} + t_0 b_{i+1}^{(k-1)} \text{ and } c = (1 - t_0)a + t_0 b. \tag{11.4}$$

Another interesting property of this representation is related to Descartes' Law of signs. The definition of Descartes' Law for a sequence of coefficients

$$\mathbf{b}_k = b_i | i = 1, ..., k$$

is defined recursively:

$$V(\mathbf{b}_{k+1}) = V(\mathbf{b}_k) + \begin{cases} 1, \text{ if } b_i b_{i+1} < 0 \\ 0, \text{ else} \end{cases} \tag{11.5}$$

With this definition, we have:

Theorem 1. *Given a polynomial* $f(x) = \sum_i^n b_i B_i^d(x; a, b)$, *the number* N *of real roots of* f *on* $]a, b[$ *is less than or equal to* $V(\mathbf{b})$, *where* $\mathbf{b} = (b_i), i = 1, ..., n$ *and* $N \equiv V(\mathbf{b}) \mod 2$.

The theorem 1 enables a simple yet efficient test of the existence of real roots in a given domain. This test is essential to our algorithm, as it serves as a key criterion to classify whether a domain has certified topology, without actually computing the curve. This allow the our algorithm to execute in reasonably short time, as demonstrated in our experiments.

11.2.2 Generalization to the multivariate case

The univariate Bernstein basis representation can be generalized to multivariate ones. Briefly speaking, we can rewrite the definition (Eq. (11.1)) in the form of tensor products. Suppose for $\mathbf{x} = (x_0, ..., x_{n-1}) \in \mathbb{R}^n$, $f = (\mathbf{x}) \in \mathbb{K}[\mathbf{x}]$ having the maximum degree $\mathbf{d} = (d_0, ..., d_{n-1})$ has the form:

$$f(\mathbf{x}) = \sum_{k_0=0}^{d_0} ... \sum_{k_n=0}^{d_n} b_{k_1,...,k_n} B_{k_0}^{d_0}(x_0)...B_{k_n}^{d_0}(x_n) \tag{11.6}$$

For a polynomial of n variables, the coefficients can be viewed as a tensor of dimension n.

The de Casteljau subdivision for the multivariate case proceeds similarly to the univariate one, since the subdivision can be done independently with regards to a particular variable x_i.

Based on these properties, a subdivision solver which can be seen as an improvement of the *Interval Projected Polyhedron* algorithm in [18], is described in [14]. It uses the following operations: The multivariate functions to be solved are enclosed in-between two univariate functions, for each variable. For this purpose, the Bernstein control points of the functions are projected in each direction and the upper and lower envelop are used to define these enveloping univariate polynomials. A lower and upper approximation of the roots of these univariate polynomials are used to reduce the domain. If the reduction is not sufficient, the domain is split. These reduction operations are improved by pre-conditioning steps. See [14] for more details.

11.3 Algorithmic ingredients

We consider the problem of computing the topology of the curve, denoted hereafter as $\mathcal{C}$, resulting from the intersection of two known algebraic surfaces, namely, $f(\mathbf{x}) = 0$ and $g(\mathbf{x}) = 0$ defined in $\mathbb{R}^3$, with $f, g \in \mathbb{R}[x, y, z]$. Our discussion is confined to the case where f and g has no common divisor other than 1, so that their intersection has dimension of 1. We assume moreover that (f, g) is radical or equivalently that the resultant of $f(x, y, z), g(x, y, z)$ with respect to z after a generic change of coordinates, is square free.

11.3.1 Tangent vector field

The tangent vector on $\mathcal{C}$ serves as the key to our analysis of topology of the curve. It serves as an important indicator of topological feature of $\mathcal{C}$. While it is computationally prohibitive to compute the tangent vector at each point on $\mathcal{C}$, we can reach some useful conclusion about the topology of the curve by looking into the tangent vector field defined below:

$$\mathbf{t} = \mathbf{t}_x(\mathbf{x})\mathbf{e}_x + \mathbf{t}_y(\mathbf{x})\mathbf{e}_y + \mathbf{t}_z(\mathbf{x})\,\mathbf{e}_z = \bigtriangledown f \wedge \bigtriangledown g = \begin{vmatrix} \mathbf{e}_x & \mathbf{e}_y & \mathbf{e}_z \\ \partial_x f & \partial_y f & \partial_z f \\ \partial_x g & \partial_y g & \partial_z g \end{vmatrix} \qquad (11.7)$$

where $\mathbf{e}_x$, $\mathbf{e}_y$ and $\mathbf{e}_z$ are the unit vectors along the principle axis x, y and z, respectively; $\mathbf{t}_x$, $\mathbf{t}_y$ and $\mathbf{t}_z$ are functions of $\mathbf{x} = (x, y, z)$.

Singularities on the curve can be easily characterized, as $\mathbf{t}$ vanishes at those points. In [8], the author also tried to localize the point having a tangent parallel to a virtual sweeping plane. They are connected together with the singularities to form the final topological graph. In order to do this, the whole curve is projected onto some principle projection planes. However, the projected planar curve in many cases has a very different topology as $\mathcal{C}$. In our proposed algorithm, we exploit the subdivision along all three principle axes simultaneously and the critical events are either reduced to regular case (such as for tangents) or localized (such as for intersections). The topology graph can be built without explicitly computing the exact position of the singularities.

11.3.2 Regularity test

In this section, we are going now to describe how to detect boxes, for which the topology of the curve can be determined. We will use the following notions:

Definition 2. *We say that a curve $\mathcal{C} \in \mathbb{R}^n$ is regular in a compact domain $D \subset \mathbb{R}^n$, if its topology is uniquely determined from its intersection with the boundary D.*

The aim of the method is to give a simple criterion for the regularity of a curve in a box.

To form the topological graph for this domain, we only need to compute the intersections between the curve and the boundary of this domain, and there exists a unique graph to link these intersections so that this graph complies to the true topology of the original curve.

2D case:

For 2D planar algebraic curve $\mathcal{C}$ defined by a polynomial equation $f(x, y) = 0$, and denoting the partial derivative of f w.r.t x by $\partial_x f$, we have the following direct property:

Proposition 3. *If $\partial_y f(x, y) \neq 0$ (resp. $\partial_x f(x, y) \neq 0$) in a domain $D = [a_0, b_0] \times [a_1, b_1] \subset \mathbb{R}^2$, the curve C is regular on D.*

Proof. Suppose that $\partial_y f(x, y) \neq 0$ in D. Then C is smooth, since its normal vector is defined everywhere, and has no vertical tangents in D. By the implicit function theorem, the connected components of $C \cap \overset{\circ}{D}$ are the graph of functions of the form $y = \varphi(x)$. The closure of such a connected component is called hereafter a branch of C in D. As $\partial_y f(x, y) \neq 0$ in D, for a given $x \in [a_0, b_0]$ there is at most one branch of C in D above x. Consequently, the connected components of $C \cap \overset{\circ}{D}$ project bijectively onto non-overlapping open intervals of $[a_0, b_0]$.

Moreover, as there is no vertical tangent, each of these branches starts and ends at a point on the border ∂D of D. Notice that two branches may share a starting or ending point, when the curve is tangent (with even multiplicity) to ∂D.

Thus, computing the points of $C \cap \partial D$, repeating a point if its multiplicity is even, sorting them by lexicographic order such that $x > y$ ($(x_0, y_0) > (x_1, y_1)$ if $x_0 > x_1$ or $x_0 = x_1$ and $y_0 > y_1$), we obtain a sequence of points $p_1, p_2, \ldots, p_{2s-1}, p_{2s}$ such that the curve C in D is isotopic to the union of the non-intersecting segments $[p_1, p_2], \ldots, [p_{2s-1}, p_{2s}]$. In other words, the topology of C is uniquely determined from its intersection points with ∂D and C is regular on D. $\square$

If $\partial_y f \neq 0$ on D (resp. $\partial_x f \neq 0$), we will say that C is x-regular (resp. y-regular). A sufficient condition for f to be x-regular (respectively y-regular) is that the Bernstein coefficients of the first derivative of f against y (respectively x) maintains a constant sign (see also [1]). By Descartes' law, this statement implies that the sign variation in this direction should be at most 1.

To put it in another way, by solely studying the sign variations of the tangential gradient vector of the curve (represented in Bernstein basis), i.e. $(\partial_y f(\mathbf{x}), -\partial_x f(\mathbf{x}))$, we are able to detect when the curve is regular on D and to determine uniquely the topological graph.

3D case:

The 2D approach can be generalized to the 3D case where the tangential gradient vector of the curve C defined by the intersection of two algebraic surfaces, namely $f(x, y, z) = 0$ and $g(x, y, z) = 0$, is given by $\mathbf{t} = \bigtriangledown(f) \wedge \bigtriangledown(g)$ (see Eh. (11.7)). Similar to the 2D case, we can represent each component of $\mathbf{t}$ in the Bernstein basis for a given domain (in cube shape) $D = [a_0, b_0] \times [a_1, b_1] \times [a_2, b_2]$. The sign change of the resulting Bernstein coefficients enables a simple regularity test with minimal computation effort.

We describe a first and simple regularity criterion:

Proposition 4. *The 3D spatial curve C defined by $f = 0$ and $g = 0$ is regular on D, if*

- $t_x(\mathbf{x}) \neq 0$ *on D, and*
- $\partial_y h \neq 0$ *(or $\partial_z h \neq 0$) on D, for $h = f$ or $h = g$.*

Proof. Suppose that $\mathbf{t}_x(\mathbf{x}) \neq 0$ and $\partial_z(f) \neq 0$ on D. It implies that $\mathcal{C}$ is smooth in D. Consider two branches of $\mathcal{C}$ in D and project them by π_z onto a (x, y)-plane. Their projection cannot intersect at an interior point. Otherwise, there would be two points $p_1, p_2 \in D$, such that $f(p_1) = 0, f(p_2) = 0$ and $\pi_z(p_1) = \pi_z(p_2)$, which implies that $\partial_z f(p)$ vanishes for an intermediate point $\in]p_1, p_2[$ in D. This is impossible by hypothesis. Consequently, the branches of $\mathcal{C}$ project bijectively onto the branches of $\pi_z(\mathcal{C})$. Their tangent vector is the projection $(\mathbf{t}_x(\mathbf{x}), \mathbf{t}_y(\mathbf{x}))$ of the tangent vector of $\mathcal{C}$. By proposition 3, $\pi_z(\mathcal{C})$ is regular, so that the topology of $\pi_z(\mathcal{C})$, and thus of $\mathcal{C}$, is uniquely determined by the intersection points of $\mathcal{C}$ with the border of D.   □

A similar criterion applies by symmetry, exchanging the roles of the x, y, z coordinates.

Let us give now a finer regularity criterion, which is computationally less expensive:

Proposition 5. *If $\mathcal{C}$ is smooth in D and if for all $x_0 \in \mathbb{R}$, the plane $\mathbf{x} = x_0$ plane has at most one intersection point with the curve $\mathcal{C}$ in D, then $\mathcal{C}$ is regular on D.*

Proof. Consider the projection $\pi_z(\mathcal{C})$ of the curve $\mathcal{C}$ in D along the z direction. Then the components of $\mathcal{C}$ in D projects bijectively on the (y, z) plane. Otherwise, there exist two points p_0 and p_1 lying on $\mathcal{C}$ such that $\pi_z(p_0) = \pi_z(p_1) = (x_0, y_0)$, then p_0 and p_1 belong to $x = x_0$ which are functions of the form $y = \Phi(x)$. Otherwise, there exist two points on $\pi_z(\mathcal{C})$ and (and on $\mathcal{C} \cap D$) with the same x-coordinate. Consequently, for $x \in [a_0, b_0]$ there is at most one branch of $\pi_z(\mathcal{C})$ in D above x, and the connected components of $\mathcal{C} \cap \overset{\circ}{D}$ project bijectively onto non-overlapping open intervals of $[a_0, b_0]$ as $\pi_z(\mathcal{C})$ does. We conclude as in the 2D case (proposition 3), by sorting the points of $\mathcal{C} \cap \partial D$ according to their x-coordinates, and by gathering them by consecutive pairs corresponding to the starting and ending points of branches of $\mathcal{C} \cap D$.   □

Proposition 6. *The 3D spatial curve $\mathcal{C}$ defined by $f = 0$ and $g = 0$ is regular on D, if*

- $t_x(\mathbf{x}) \neq 0$ *on D, and*
- $\partial_y h \neq 0$ *on z-faces, and $\partial_z h \neq 0$ and its has the same sign on both y-faces of D, for $h = f$ or $h = g$.*

Proof. Let us fix $x_0 \in [a_0, b_0]$ where $D = [a_0, b_0] \times [a_1, b_1] \times [a_2, b_2]$, let $U = \{x_0\} \times [a_1, b_1] \times [a_2, b_2]$ and let $\Phi_{x_0} : (x_0, y, z) \in U \mapsto (f(x_0, y, z), g(x_0, y, z))$. We are going to prove that under our hypothesis, Φ_{x_0} is injective. The Jacobian $\mathbf{t}_x(x_0, y, z)$ of Φ_{x_0} does not vanish on U, so that Φ_{x_0} is locally injective. We consider the level-set $f(\mathbf{x}) = f_0$ for some $f_0 \in f(U)$. It cannot contain a closed loop in U, otherwise we would have $(\partial_y f, \partial_z f) = 0$ (and thus $\mathbf{t}_x = 0$) in $U \subset D$. We deduce that each connected component of $f(\mathbf{x}) = f_0$ in U intersects ∂U in two points.

Now suppose that Φ_{x_0} is not injective on U, so that we have two points $p_1, p_2 \in U$ such that $\Phi_{x_0}(p_1) = \Phi_{x_0}(p_2)$.

If p_1 and p_2 are on the same connected component of the level set $f(\mathbf{x}) = f_0$ (where $f_0 = f(p_1) = f(p_2)$) in U, then g reaches the same value at p_1 and p_2 on this level set, so that by Role's theorem, there exists a point $p \in U$ in-between p_1 and p_2, such that $\mathrm{Jac}(\Phi_{x_0})(p) = \mathbf{t}_x(p) = 0$. By hypothesis, this is impossible.

Thus p_1 and p_2 belongs to two different connected components of $f(\mathbf{x}) = f_0$ in U. Consequently the value f_0 is reached at 4 distinct points of ∂U, which implies that f has at least 4 extrema on ∂U.

Now note that up to a change of variable $z = a_2 - z$, we can assume that $\partial_z f > 0$ on both $y = a_1, y = b_1$ faces. Then if $\partial_y f < 0$ on $z = a_2$, we have $f(x_0, a_1, b_2) > f(x_0, a_1, a_2) > f(x_0, b_1, a_2)$ and (a_2, a_3) is not a local extrema. Otherwise $\partial_y f > 0$ and (b_1, a_2) is not a local extrema. In both cases, we do not have 4 extrema, which proves that φ_{x_0} is injective and that the intersection of $\mathcal{C}$ with the plane $x = x_0$ in D is at most one point. So by proposition 5, we deduce that $\mathcal{C}$ is regular in D. $\qquad\qquad\square$

For more details on the injectivity properties, see [16]. Here also, a similar criterion applies by symmetry, exchanging the roles of the x, y, z coordinates.

If one of these criteria applies with $\mathbf{t}_i(x) \neq 0$ on D (for $i = x, y, z$), we will say that $\mathcal{C}$ is i-regular on D.

From a practical point of view, the test that $\mathbf{t}_i(x) \neq 0$ or $\partial_i(h)$ for $i = x, y$ or z, $h = f$ or g, is replaced by the stronger condition that their coefficients on the Bernstein basis of D have a constant sign, which is straightforward to check. Similarly, such a property on the faces of D is also direct, since the coefficients of a polynomial, with a minimal (resp. maximal) x-indices (resp. y-indices, z-indices) are its Bernstein coefficients on the corresponding face.

In addition to these tests, we also test whether both surfaces penetrate the cell, since a point on the curve must lie on both surfaces. This test could be done by looking at the sign change of the Bernstein coefficients of the surfaces with regards to that cell. If no sign change occurs, we can rule out the possibility that the cell contains any portion of the curve $\mathcal{C}$, hence terminate the subdivision early. In this case, we will also say that the cell is regular.

The regularity criterion is sufficient for us to uniquely construct the topological graph g of $\mathcal{C}$ within D. Without loss of generality, we suppose that the curve $\mathcal{C}$ is x-regular in D. Hence, there is no singularity of $\mathcal{C}$ in D. Furthermore, this also guarantees that there is no 'turning-back' of the curve tangent along x-direction, so the mapping of $\mathcal{C}$ onto the x axis is injective. Intuitively, the mapped curve should be a series of non-overlapping line segments, of which the ends correspond to the intersections between the curve $\mathcal{C}$ and the cell, and such mapping is injective.

This property leads to a unique way to connect those intersection points, once they are computed (see section 11.3.3), in order to obtain a graph representing the topology of $\mathcal{C}$. Here is how this graph is computed in practice: suppose $\mathcal{C}$ is i-regular in the domain D, and that we have computed the set of intersection points $V = \{\mathbf{v}_j\}$ of the curve with the boundary of D. First, we sort the elements in V comparing vectors by their i-th coordinates. Assuming the sorted points $\mathbf{v}_j$ are indexed by $j = 0, 1, 2, ...$, we form the edges $\mathbf{v}_k, \mathbf{v}_{k+1}$, for $k = 0, 2, 4, ...$

However, a special case has to be taken into account, that is when $\mathbf{v}_j$ has a multiplicity $m_i > 1$, for instance, when $\mathcal{C}$ is tangent to the bounding domain D at $\mathbf{v}_i$. In this case, we can treat $\mathbf{v}_j$ conceptually as a multiple point which plays the role of m_i points. In this way, we proceed the connecting process in the same manner as we do for the general case. To determine the multiplicity of a point $\mathbf{v}_j$, we only have to evaluate the derivatives of $\mathcal{C}$ at this point.

11.3.3 Hierarchical subdivision

We adopted a hierarchical octree to partition the $\mathbb{R}^3$ space, for several reasons:

- each cell of the octree is equivalent to a cube-shaped domain D; which stores the coefficients of the polynomials in the Bernstein basis of the corresponding domain.
- we can take cares of faces shared by cells, to minimize the number of calls to solvers;
- the hierarchical structure of octree allows us to terminate (stop further subdivision) early when a cell is deemed regular or irrelevant.

We begin by setting a initial bounding domain D_0 to a root cell. A cell is subdivided if the curve $\mathcal{C}$ defined in the correspondent domain fails the regularity test. For each subdivision, we result in several smaller domains in form of sub-cells. For each of them, we repeat the regularity test and, if necessary, further subdivides. The subdivision of a cell will terminate either when the curve within is deemed regular, or the size of the cell is beyond a predefined precision ϵ.

There are several techniques to save computation efforts. As the sub-cells share certain faces with their parent cell, the earlier computed intersections on the parent cell's faces are inherited directly by the sub-cells. In addition, sub-cells split from the same parent cell do share some faces as well. Once again, the shared faces should be computed exactly once.

Once a new face is introduced in the octree decomposition, the bivariate solver described in section 11.2.2, is called directly with the Bernstein coefficients of the polynomials on this face. The points we found are shared by neighbor cells, connected to this face in the octree.

11.3.4 Symbolic-numeric approach

Some geometric operations such as computing the self-intersection curve or the ridge curve of a parameterized surface leads to the computation of implicit curves of high degree with coefficients of large size. This is either due to projection techniques (see [9]), or to their definition through composed operations (see [3]). In order to be able to handle such curve, the main difficulty is to control the result, using approximate computation, since exact computation though possible, would be prohibitive. We describe here the symbolic-numeric approach that we have developed for this purpose.

We assume that the input equations are given with exact (large) rational (or integer) numbers (even if the input is given with floating point numbers, we will consider it as an exact input). In order to compute the topology of C in a domain D, we convert its representation in the Bernstein basis of D, using exact rational arithmetic.

Once this conversion is done, we normalize the equation, by dividing by the coefficient of maximal norm. For each resulting rational coefficient c, we compute the smallest interval $[\underline{c}, \overline{c}]$ represented with floating point numbers and containing c.

Then, the subdivision process is performed, using interval arithmetic. The regularity criterion, which reduces to sign evaluations, is applied on these interval coefficients. We use the following convention: a interval is < 0 (resp. > 0) if all its elements are < 0 (resp. > 0). If the interval contains 0, we say that its sign is indeterminate.

If the regularity test fails,

- either the sign of all the coefficients of the polynomial are indeterminate, and we re-convert the exact polynomial to its representation on the corresponding sub-domain and restart the approximation process.
- or we subdivide the domain, as in the usual case.

11.3.5 Outline of the algorithm

The proposed algorithm for 3D curves is outlined as follows:

We do not describe the algorithm for 2D curves, which is basically a specialization of this one.

11.4 Experiments

Our proposed algorithm is implemented as a part of SYNAPS (SYmbolic Numeric APplicationS) library[2]. The experiments have been carried out on a 3.4GHz PC, under Linux.

11.4.1 Planar curves of high degree with large coefficients

In this section, we report on the application of the 2d algorithm, in the case of large integer coefficients. The first example is about ridge curve. Ridge curves correspond to local extrema of curvature taken in the principal direction of the surface, after some algebraic manipulations they can be obtained as implicit curve (see [3]). See also [19] and [20] for other related approaches. In the example it corresponds to a bicubic surface, the input polynomial is of total degree 84, of multidegree $(43, 43)$ with 1907 monomials. The coefficients are integers encoded on at most 65 bits. For the precision $\epsilon = 10^{-3}$ which controls the singularity localization, it takes 30 seconds. The topology is certified except in tiny boxes (which contains the singularity

---

[2] http://www-sop.inria.fr/galaad/software/synaps/

Computing the topology of the curve $\mathcal{C}$:

INPUT: f($\mathbf{x}$) and g($\mathbf{x}$) polynomials $\in \mathbb{Q}[x, y, z]$, a tolerance ϵ and a list of bounding domain $D_0 \leftarrow [a_0, b_0] \times [a_1, b_1] \times [a_2, b_2]$ $(a_{i,b_i} \in \mathbb{R})$.

- Step 0: (initialization step) domain list $\mathcal{D} \leftarrow D_0$; vertex list $V \leftarrow$ NIL; connectivity list $E \leftarrow$ NIL;
- Step 1: compute $\mathbf{t} \leftarrow \bigtriangledown(f) \wedge \bigtriangledown(g)$ given by Eq. (11.7);
- Step 2: convert f, g and $\mathbf{t}$ into Bernstein basis representation;
- Step 3: while $\mathcal{D}$ is not empty, pick a D in $\mathcal{D}$:
  - Step 3.1: compute V the set of intersection points between the boundary of the domain D and the curve $\mathcal{C}$;
  - Step 3.2: if the size of D is larger than ϵ:
    - if the curve $\mathcal{C}$ within the domain D is regular (see section 11.3.2):
      - sort and connect the points $\mathbf{v} \in V$; the connectivities are stored in E;
    - else if the domain D is not regular:
      - subdivide D and append the subdivided domains into the domain list D
  - else if the size of D is not larger than ϵ:
    - add the domain D as a 'box' vertex into V;
    - this vertex is connected with all intersections $\mathbf{v} \in V$ of D; these connectivities are also appended to E;
  - Step 3.4: remove D from $\mathcal{D}$ and repeat Step 3;

OUTPUT: The graph represented by a set of vertices V, which are either 3D points or boxes (with size less than ϵ) bounding the singularities, and a set of connections E that are representing the edges of the resulting graph.

points). Notice that a pure algebraic approach, exploiting the specificity of problem and with a very efficient Gröbner engine takes about 10 minutes to certify the topology (see [3]).

The second example is a projection of a self-intersection curve of bicubic patch, computed by resultant techniques (see [9]). The input polynomial is of total degree 76, of multidegree $(44, 44)$ with 1905 monomials. The coefficients are integers of at most 288 bits. It takes 5 seconds, for this example with the same precision $\epsilon = 10^{-3}$.

11.4.2 Intersection curves of implicit surfaces

This set of examples are from [8]. The computational time accompanied is measured up to milliseconds (see Fig. 11.2):

1) $f(\mathbf{x}) = 0.85934x^2 + 0.259387xy + 0.880419y^2 + 0.524937xz - 0.484008yz + 0.510242z^2 - 1$
   $g(\mathbf{x}) = 0.95309x^2 + 0.303149xy + 0.510242y^2 - 0.200075xz + 0.64647yz + 0.786669z^2 - 1$
   time: 80 msec
2) $f(\mathbf{x}) = -0.125x^2 - 0.0583493xy + 0.493569y^2 + 0.966682xz - 1.5073yz - 0.368569z^2 - 0.865971x - 0.433067y - 0.250095z$

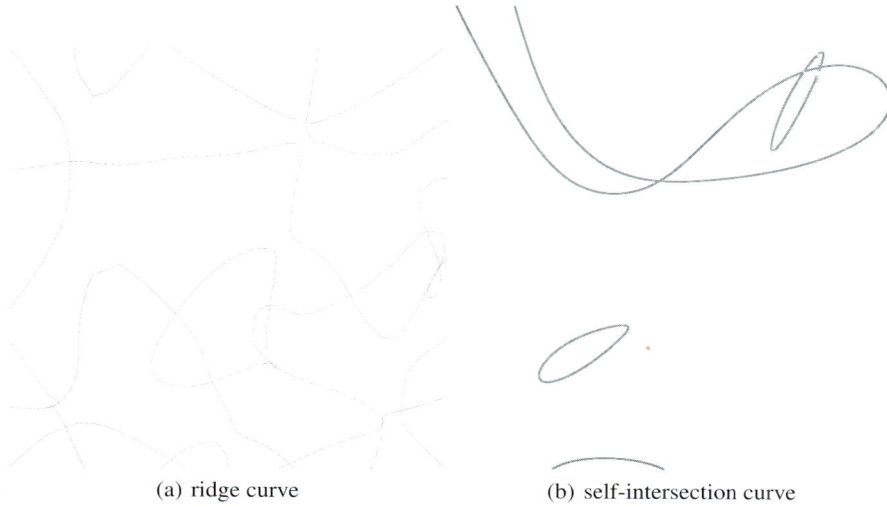

(a) ridge curve   (b) self-intersection curve

Fig. 11.1. Topological descriptions of high complexity curves

$g(\mathbf{x}) = x^2 + y^2 + z^2 - 2$
time: 20 msec
3) $f(\mathbf{x}) = 2x^2 + y^2 + z^2 - 4$
$g(\mathbf{x}) = x^2 + 2xy + y^2 - 2yz - 2z^2 + 2zx$
time: 30 msec
4) $f(\mathbf{x}) = x^4 + y^4 + 2x^2y^2 + 2x^2 + 2y^2 - x - y - z$
$g(\mathbf{x}) = x^4 + 2x^2y^2 + y^4 + 3x^2y - y^3 + z^2$
time: 130 msec

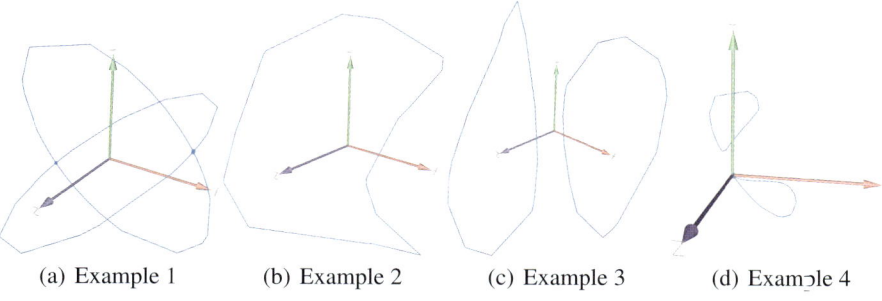

(a) Example 1   (b) Example 2   (c) Example 3   (d) Example 4

Fig. 11.2. Topological descriptions of the intersection curve for 4 pairs of low-order algebraic surfaces.

11.4.3 Silhouette curves of implicit surfaces

The following samples are taken from *http://www-sop.inria.fr/galaad/surfaces/*. We intersect the surface with its polar variety in one direction (here the x direction). In other words, we intersect the surface with the surface defined by one of its first order derivative (here $\partial_x f$), to extract its silhouette. The surfaces that we used are called respectively *Tetrahedral, Q3, Q1* and *Barth Sextic* (see Fig.11.3):

5) $f(\mathbf{x}) = x^4 + 2x^2y^2 + 2x^2z^2 + y^4 + 2y^2z^2 + z^4 + 8xyz - 10x^2 - 10y^2 - 10z^2 + 25$
    $g(\mathbf{x}) = 4x^3 + 4xy^2 + 4xz^2 + 8yz - 20x$
    time: 510 msec

6) $f(\mathbf{x}) = 5.229914547374508y^2z^2 + 3.597883597883598x^2y^2 + y^4 + z^4 - x^4 - 19.49816368932737xyz + 5.229914547374508x^2 - 7.43880040039534y^2$
    $g(\mathbf{x}) = -3.59788359788359z^2 + 7.43880040039534z^2x^2 - 110.45982909yz^2 + 7.195767196x^2y + 4y^3 - 19.49816368932737xz - 14.87760080y$
    time: 330 msec

7) $f(\mathbf{x}) = x^4 + y^4 + z^4 - 4x^2 - 4y^2z^2 - 4y^2 - 4z^2x^2 - 4z^2 - 4x^2y^2 + 20.7846xyz + 1$
    $g(\mathbf{x}) = 4x^3 - 8x - 8xz^2 - 8xy^2 + 20.7846yz$
    time: 730 msec

8) $f(\mathbf{x}) = 67.77708776x^2y^2z^2 - 27.41640789x^4y^2 - 27.41640789x^2z^4$
    $+ 10.47213596x^4z^2 - 27.41640789y^4z^2 + 10.47213596y^4x^2 + 10.47213596y^2z^4$
    $- 4.236067978x^4 - 8.472135956x^2y^2 - 8.472135956x^2z^2 + 8.472135956x^2$
    $- 4.236067978y^4 - 8.472135956y^2z^2 + 8.472135956y^2 - 4.236067978z^4$
    $+ 8.472135956z^2 - 4.236067978$
    $g(\mathbf{x}) = 135.5541755xy^2z^2 - 109.6656316x^3y^2 - 54.83281578xz^4$
    $+ 41.88854384x^3z^2 + 20.94427192y^4x - 16.94427191x^3 - 16.94427191xy^2$
    $- 16.94427191xz^2 + 16.94427191x$
    time: 4010 msec

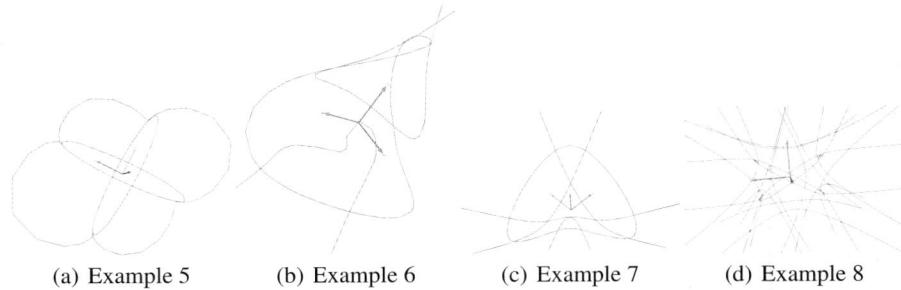

(a) Example 5   (b) Example 6   (c) Example 7   (d) Example 8

Fig. 11.3. Topological descriptions of the silhouette curves of algebraic surfaces.

11.5 Discussion

The algorithm proposed in this paper offers a generic method for computing the topological graph of spatial curves resulting from the intersection of two algebraic surfaces. As demonstrated in the experiments, it is rather robust despite the increase of the complexity of the curve.

The major weakness of this approach, however, is that certain apparently simply situation could result in a lot of subdivisions, such as the curve with parallel structures which are very close to each other.

As specified, in this method, the singular points are only isolated into small boxes of size ϵ, and we do not certify the connection of the branches at these points. An additional work would be necessary, to certify the singularity type. We are currently investigating on this problem.

Another weakness of this approach is the memory consumption (as the storage requirement is approximately cubic to the depth of the subdivision). This problem however, can be mitigated by a divide-and-conquer approach and pure programming techniques.

Acknowledgements:

We would like to thanks J. Wintz, for his very nice software AXEL[3] for the visualisation and manipulation of algebraic objects, that we used to produce the pictures of the curves. We acknowledge the partial support of Aim@Shape (IST NoE 506766) and ACS (IST Fet Open 006413).

References

1. L. Alberti, G. Comte, and B. Mourrain. Meshing implicit algebraic surfaces: the smooth case. In L.L. Schumaker M. Maehlen, K. Morken, editor, *Mathematical Methods for Curves and Surfaces: Tromso'04*, pages 11–26. Nashboro, 2005.
2. J. Gerardo Alcázar and J. Rafael Sendra. Computing the topology of real algebraic space curves. *J. Symbolic Comput.*, 39:719–744, 2005.
3. F. Cazals, J.-C. Faugère, M. Pouget, and F. Rouillier. Topologically certified approximation of umbilics and ridges on polynomial parametric surface. Technical Report 5674, INRIA Sophia-Antipolis, 2005.
4. G. Elber and M.-S Kim. Geometric constraint solver using multivariate rational spline functions. In *Proc. of 6th ACM Symposium on Solid Modelling and Applications*, pages 1–10. ACM Press, 2001.
5. G. Farin. *Curves and surfaces for computer aided geometric design : a practical guide.* Comp. science and sci. computing. Acad. Press, 1990.
6. G. Farin. *Curves and Surfaces for Computer Aided Geometric Design: A Practical Guide, 3rd Ed.* Academic Press, 1993.
7. T. Grandine and F. Klein. A new approach to the surface intersection problem. *Computer Aided Geometric Design*, 14(2):111–134, 1997.

---

[3] http://www-sop.inria.fr/galaad/software/axel/

8. G. Gatellier, A. Labrouzy, B. Mourrain, and J.-P. Técourt. *Computing the topology of 3-dimensional algebraic curves*, pages 27–44. Springer-Verlag, 2005.

9. A. Galligo and J.P. Pavone. Self-intersections of a Bézier bicubic surface. In M. Kauers, editor, *Proc. Intern. Symp. on Symbolic and Algebraic Computation*, pages 148–155. New-York, ACM Press., 2005.

10. L. González-Vega and I. Necula. Efficient topology determination of implicitly defined algebraic plane curves. *Computer Aided Geometric Design*, 19(9):719–743, 2002.

11. J. Hass, R. T. Farouki, C. Y. Han, X. Song, and T. W. Sederberg. Guaranteed Consistency of Surface Intersections and Trimmed Surfaces Using a Coupled Topology Resolution and Domain Decomposition Scheme. *Advances in Computational Mathematics*, 2005. To appear.

12. Seong Joon-Kyung, Elber Gershon, and Kim Myung-Soo. Contouring 1- and 2-Manifolds in Arbitrary Dimensions. In *SMI'05*, pages 218–227, 2005.

13. J. Keyser, T. Culver, D. Manocha, and S. Krishnan. Efficient and exact manipulation of algebraic points and curves. *Computer-Aided Design*, 32(11):649–662, 2000.

14. B. Mourrain and J.-P. Pavone. Subdivision methods for solving polynomial equations. Technical Report 5658, INRIA Sophia-Antipolis, 2005.

15. B. Mourrain, F. Rouillier, and M.-F. Roy. *Bernstein's basis and real root isolation*, pages 459–478. Mathematical Sciences Research Institute Publications. Cambridge University Press, 2005.

16. J.P. Pavone. *Auto-intersection de surfaces pamatrées réelles*. PhD thesis, Université de Nice Sophia-Antipolis, 2004.

17. T. Sederberg. Algorithm for algebraic curve intersection. *Computer-Aided Design*, 21:547–554, 1989.

18. E. C. Sherbrooke and N. M. Patrikalakis. Computation of the solutions of nonlinear polynomial systems. *Comput. Aided Geom. Design*, 10(5):379–405, 1993.

19. J.-Ph. Thirion and A. Gourdon. The 3D Marching Lines Algorithm and its Applications to Crest Lines Extraction. Technical Report 1672, INRIA, 1992.

20. J.-Ph. Thirion and A. Gourdon. Computing the Differential Characteristics of Isointensity Surfaces. *Computer Vision and Image Understanding*, 61(2):190–202, 1995.

Approximate Implicitization of Space Curves and of Surfaces of Revolution

Mohamed Shalaby and Bert Jüttler

Institute of Applied Geometry,
Johannes Kepler University, Linz, Austria
bert.juettler@jku.at

Summary. We present techniques for creating an approximate implicit representation of space curves and of surfaces of revolution. In both cases, the proposed techniques reduce the problem to that of implicitization of planar curves. For space curves, which are described as the intersection of two implicitly defined surfaces, we show how to generate an approximately orthogonalized implicit representation. In the case of surfaces of revolution, we address the problem of avoiding unwanted branches and singular points in the region of interest.

12.1 Introduction

Traditionally, most CAD (Computer Aided Design) systems rely on piecewise rational parametric representations, such as NURBS (Non–Uniform Rational B–Spline) curves and surfaces. The parametric representation offers a number of advantages, such as simple sampling techniques, which can be used for quickly generating an approximating triangulation for visualization. On the other hand, the use of implicitly defined curves/surfaces also offers a number of advantages, e.g., for solving intersection problems, or for visualization via ray–tracing.

In order to exploit the potential benefits of using the implicit representation of curves and surfaces, methods for conversion from parametric to implicit form (implicitization) are needed. As an alternative to exact methods, such as resultants, Gröbner bases, moving curves and surfaces, etc. [2, 4, 5, 8, 14], a number of approximate techniques have emerged [3, 7, 10, 11]. As demonstrated in the frame of the European GAIA II project [6, 15, 17], these techniques are well suited to deal with general free–form curve and surface data arising in an industrial environment.

On the other hand, CAD objects typically involve many special curves and surfaces, such as natural quadrics, sweep surfaces, surfaces of revolution, etc. While implicit representations of simple surfaces are readily available, this paper studies approximate approximation of two special objects, namely space curves and surfaces of revolution. Space curves arise frequently in geometric modeling. An implicit representation of a space curve is given by the intersection of two implicitly defined surfaces. A surface of revolution is created by rotating a 2D profile curve about an axis in space. Rotation is one of the standard geometric operations defined in any CAD/CAM interface.

This paper presents techniques for approximate implicitization of space curves and of surfaces of revolution, which are based on the (approximate) implicitization of planar curves. The proposed techniques are fully general in the sense that they can be combined with any (exact or approximate) implicitization method for planar curves. For creating the examples shown in this paper, we used a technique for simultaneous approximation of points and associated normal vectors [10, 11, 16].

This paper is organized as follows. First we summarize the approximate implicitization method for planar curves. Section 12.3 presents techniques for approximate implicitization of space curves, first as the intersection of two general cylinders, and later as the intersection of two general surfaces which intersect approximately orthogonal. Representing the space curve by two 'orthogonal' surfaces provides a more robust definition for the curve. Finally, in Section 12.4, two methods for approximate implicitization of surfaces of revolution are presented. It is shown that – in many cases – only approximate implicitization is capable of producing an implicit representation that is free of unwanted branches and singularities.

12.2 Simultaneous approximation of points and normals

For the sake completeness, we give a short description of the approximate implicitization method presented in [10] (see also [11] for the case of surfaces). This method is characterized by the simultaneous approximation of sampled point data $\mathbf{p}_i = (x_i, y_i)$, $i \in \mathcal{I} = \{1, \ldots, N\}$, and estimated unit normals $\mathbf{n}_i$ at these points. The method consists of three main steps:

- *Step 1 – Preprocessing:* If no other information is available (e.g., from a given parametric or procedural description of the curve), then each unit normal vector $\mathbf{n}_i$ is estimated from the nearest neighbors of the point $\mathbf{p}_i$. A consistent orientation of the normals is achieved by a region–growing–type process. If the data have been sampled from a curve with singularities, then it may be necessary to organize the data into several segments, see [16] for details.
- *Step 2 – Fitting:* We generate an approximate implicit representation of the form

$$f(\mathbf{x}) = \sum_{j \in \mathcal{J}} c_j \, \varphi_j(\mathbf{x}) \qquad (12.1)$$

with certain coefficients $c_j \in \mathbb{R}$, finite index set $\mathcal{J}$ and suitable basis functions φ_j. For instance, one may choose tensor–product B-splines with respect to suitable knot sequences, or Bernstein polynomials with respect to a triangle containing the data.

The coefficients of f are obtained as the minimum of

$$\sum_{i \in \mathcal{I}} f(\mathbf{p}_i)^2 + w_1 \|\nabla f(\mathbf{p}_i) - \mathbf{n}_i\|^2 + w_2 \, T, \qquad (12.2)$$

where w_1 and w_2 are positive weights satisfying $1 > w_1 \gg w_2 > 0$. The first weight controls the influence of the estimated normal vectors $\mathbf{n}_i$ to the resulting curve. As observed in our experiments, increasing the weights w_1 or w_2 can be used to 'push away' unwanted branches of the curve from the region of interest. The tension term T in (12.2) is added in order to control the shape of the resulting curve. It pulls the approximating curve towards a simpler shape. A possible quadratic tension term is

$$T = \iint\limits_{\Omega} f_{xx}^2 + 2\, f_{xy}^2 + f_{yy}^2 \;\; \mathrm{d}x\,\mathrm{d}y \qquad (12.3)$$

This choice of the tension term leads to a positive definite quadratic objective function. Consequently, the coefficients c_j are found by solving a system of linear equations. In the case of tensor–product B–splines, this system is sparse.

- *Step 3 – Iteration:* One may iterate the second step, by replacing the normals $\mathbf{n}_i$ with the gradients $\nabla f(\mathbf{p}_i)$, and re–computing the result. One the one hand, this may help to improve the result of the fitting. On the other hand, it can create problems with unwanted branches. This is described in some detail in [10].

Example 1. We illustrate the behaviour of exact and approximate implicitization by an example. Figure 12.1 shows the results (algebraic curves of order 4) of both methods (thin curves) for a segment of a rational planar curve of degree 4 (bold curves). The approximate implicitization produces an exact implicitization, but with additional branches and even a singular point in the region of interest. Depending on the choice of w_1, the fitting method produces implicit approximations with different level of accuracy. The weight w_1 can be used to control unwanted branches and singular points. In this example, $w_2 \approx 0$ has been chosen, and three iterations were applied to improve the result.

Remark 2. As described in [11], the distance between a parametric curve $\mathbf{p}(t)$ and its approximate implicitization can essentially be bounded by

$$\max_{t \in I}(f \circ \mathbf{p})(t) / \min_{\mathbf{x} \in \Omega} \|\nabla f(\mathbf{x})\|, \qquad (12.4)$$

where I and Ω are the domains of the parametric curve and its approximate implicitization, respectively. Upper resp. lower bounds on numerator and denominator can be obtained by using the convex–hull property of B–spline and Bézier representations. At the same time, the lower bound on $\|\nabla f(\mathbf{x})\|$ certifies the regularity of the approximate implicitization within the region of interest. If the accuracy is insufficient or the regularity is violated, then one may (semi–automatically) adjust the input parameters (number of sampled data, knots, degrees, and weights).

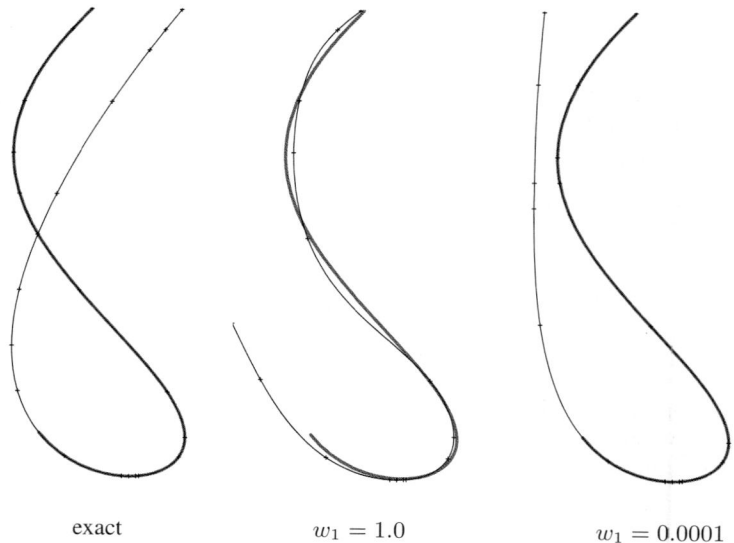

exact $w_1 = 1.0$ $w_1 = 0.0001$

Fig. 12.1. Exact (left) vs. approximate (center and right) implicitization (thin curves) of a given parametric curve (bold curves), see Example 1.

12.3 Approximate implicitization of space curves

After presenting some preliminaries, we discuss the approximate implicitization of two space curves as the intersection of two generalized cylinders and as the intersection of algebraic surfaces which are approximately orthogonal to each other.

12.3.1 Preliminaries

For any function $f : \mathbb{R}^3 \to \mathbb{R}$, the zero contour (or zero level set) $\mathcal{Z}(f)$ is the set

$$\mathcal{Z}(f) = \{\mathbf{x} \mid f(\mathbf{x}) = 0\} = f^{-1}(\{0\}) \tag{12.5}$$

A *space curve* C can be defined as the intersection curve of two zero sets of functions f and g,

$$C(f, g) = \mathcal{Z}(f) \cap \mathcal{Z}(g). \tag{12.6}$$

If both f and g can be chosen as polynomials, then $C(f, g)$ is called an *algebraic* curve. A point $\mathbf{x} \in C(f, g)$ is said to be a *regular* point of the space curve, if the gradient vectors $\nabla f(\mathbf{x})$ and $\nabla g(\mathbf{x})$ are linearly independent. The tangent vector of the space curve is then perpendicular to both gradient vectors.

The two zero contours $\mathcal{Z}(f)$ and and $\mathcal{Z}(g)$ intersect *orthogonally* along the space curve $C(f, g)$, if

$$\nabla f(\mathbf{x}) \cdot \nabla g(\mathbf{x}) = 0 \tag{12.7}$$

holds for all $\mathbf{x} \in C(f, g)$.

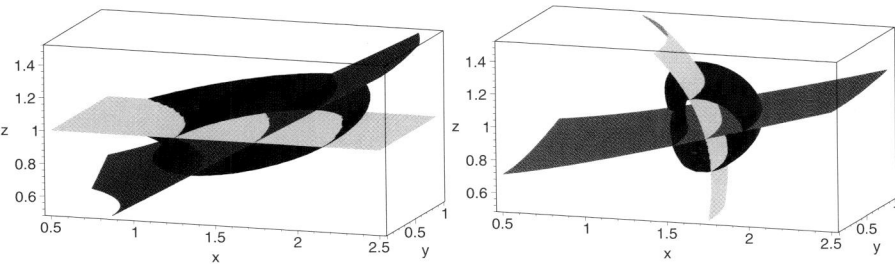

Fig. 12.2. Two surfaces, their intersection curve and a level set of the function L, see Example 3.

Representing the space curve by two surfaces which intersect orthogonally provides a more robust definition for the curve [1], since small perturbations of the defining two surfaces have less impact on the space curve. It has several additional advantages, e.g., for estimating the *Euclidean distance* of a point to the curve. As a natural generalization of the so–called *Sampson distance* $f(\mathbf{p})/\|\nabla f(\mathbf{p})\|$, see [13], this distance can be estimated as

$$L = \sqrt{\frac{f^2}{\|\nabla f\|^2} + \frac{g^2}{\|\nabla g\|^2}} \qquad (12.8)$$

In the case of two surfaces which intersect each other orthogonally, L provides a good local (i.e., in the vicinity of the intersection curve) approximation of the distance field. In a different context, orthogonalization of implicits has also been used in [12].

Example 3. Fig. 12.2 visualizes this observation. Two surfaces, their intersection curve and a level set of the function L are shown. In the case of two orthogonal surfaces (right), the level set is more similar to a pipe surface than in the general situation (left).

12.3.2 Intersection of generalized cylinders

A generalized cylinder is obtained by extruding a profile curve $\mathcal{Z}(f)$ along a straight line. If the straight line is parallel to one of the coordinate axes, say the z–axis, then the zero contour of any function of the form $(x, y, z) \rightarrow f(x, y)$ defines such a generalized cylinder.

This simple observation leads to algorithm 2 which generates an approximate implicit representation of a space curve. If step 2 uses an exact implicitization method (instead of an approximate one), then the algorithm generates an exact implicitization of the space curve.

Remark 4. Instead of the the xy and the xz plane, any two orthogonal planes can be used. Clearly, one could try to choose them such that the projection becomes as simple as possible. As an important condition, no chord of the curve should be orthogonal to one of the two planes.

Algorithm 2 Approximate implicitization by generalized cylinders

---

Input A parametric space curve $\mathbf{C}$ or a set of sampled points $\mathbf{p}_i$.
Output An implicit representation of the given space curve as the intersection of two generalized cylinders.

1: Project the parametric space curve $\mathbf{C}$ (the points $\mathbf{p}_i$) orthogonally into two orthogonal planes (e.g. xy-plane and xz-plane).
2: Apply an approximate implicitization method to the data in xy-plane and xz-plane. Let the bivariate functions $f(x, y)$ and $g(x, z)$ define the implicit curves in xy-plane and xz-plane respectively.
3: Define the two generalized cylinders by the polynomials $f(x, y)$ and $g(x, y)$ respectively.
4: Represent the curve $C(x, y, z)$ as the intersection of the two generalized cylinders $f(x, y)$ and $g(x, z)$.

Example 5. The left plot in Figure 12.4 (see page 222) shows a space curve (white) which is represented as the intersection of two generalized cylinders $\mathcal{Z}(f)$ (black) and $\mathcal{Z}(g)$ (grey), where $f = f(x, y)$ and $g = g(x, z)$.

12.3.3 Approximately orthogonal representation

Our method for generating an approximate implicitization by two approximately orthogonal surfaces is based on the following simple observation.

Lemma 6. *At all regular points* $\mathbf{x} \in C(f, g)$, *the gradients of the two functions*

$$F(\mathbf{x}) = \|\nabla f(\mathbf{x})\| \, g(\mathbf{x}) + \|\nabla g(\mathbf{x})\| \, f(\mathbf{x}) \tag{12.9}$$
$$G(\mathbf{x}) = \|\nabla f(\mathbf{x})\| \, g(\mathbf{x}) - \|\nabla g(\mathbf{x})\| \, f(\mathbf{x}) \tag{12.10}$$

are orthogonal.

This observation can be verified by a direct computation.

Remark 7. This result cannot be used at points where the two original surfaces intersect each other tangentially. In the case of two generalized cylinders produced by Algorithm 2, this happens only if the curve $\mathbf{C}$ has a tangent which lies in a plane that is perpendicular to both projection planes. One may easily choose the two projection planes such that this is not the case.

Clearly, even if the function f and g are piecewise polynomials, neither F nor G are piecewise polynomials in general. We propose to approximate them by piecewise polynomials, as follows.

The functions $\|\nabla f\|$ and $\|\nabla g\|$ depend on x, y and x, z respectively. We would like to approximate them by two piecewise polynomials $\bar{f}(x, y)$ and $\bar{g}(x, z)$ in the area of interest, which is the region near the zero contours of the functions f and g. (See [9] for more information and references on surface fitting.) The two approximating functions are to minimize

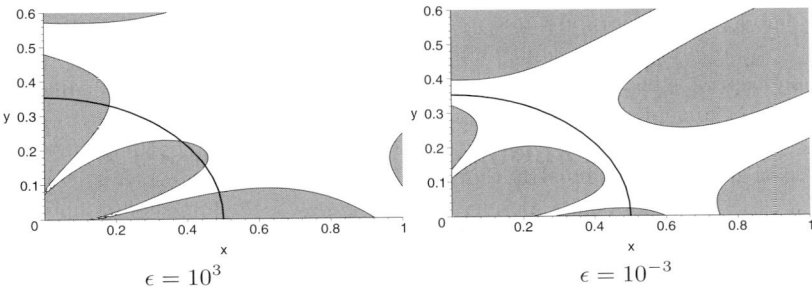

Fig. 12.3. Approximation of the scalar field $||\nabla f||$, see Example 8.

$$\iint_{\Omega'} w(f)\,(\bar{f} - ||\nabla f||)^2 \,\mathrm{d}x\,\mathrm{d}y \quad \text{and} \quad \iint_{\Omega''} w(g)\,(\bar{g} - ||\nabla g||)^2 \,\mathrm{d}x\,\mathrm{d}z \qquad (12.11)$$

where w is a suitable weight function. For instance, one may use

$$w(h) = \frac{1}{h^2 + \epsilon}, \qquad (12.12)$$

where $\epsilon > 0$ is used in order to avoid division by zero.

Note that the objective functions depend quadratically on $\bar{f}$ and $\bar{g}$. Consequently, if these approximants are represented as a linear combination of certain basis functions (such as tensor–product B-splines), similar to (12.1), then the minimizers of (12.11) can be computed by solving symmetric positive definite systems of linear equations. In the B-spline case, these systems are sparse. The coefficients of the equations have to be evaluated by numerical integration, e.g., by Gaussian quadratures.

Example 8. We consider the gradient field of $f = 4x^2 + 8y^2 - 1$ on $[0, 1] \times [0, 0.6]$ and approximate the scalar field $||\nabla f|| = 8\sqrt{x^2 + 4y^2}$ by a quadratic polynomial. For different values of ϵ we obtain different approximations. The white regions in Fig. 12.3 show where the relative error is less than 2%. For smaller values of ϵ, this region follows the elliptic arc $\mathcal{Z}(f)$, which is shown as a black line.

Algorithm 3 combines the previous algorithm with the approximation of the norms of the gradients. The degree $\deg_x(F)$ and $\deg_x(G)$ of the surfaces F and G with respect to x equals $\max(\deg_x(\bar{f}) + \deg_x(g), \deg_x(\bar{g}) + \deg_x(f))$. The degree with respect to y (and similarly for z) is $\max(\deg_x(f), \deg_x(\bar{f}))$. In order to reduce the total degree, one may consider to choose the degree of the factors $\bar{f}, \bar{g}$ as small as possible. Alternatively, one may use (tensor–product) spline functions.

Example 9. We consider a given space curve and apply the two algorithms to it. Figure 12.4 shows the approximate implicitization by two generalized cylinders (left) and by two approximately orthogonal algebraic surfaces (right). For the latter two surfaces, the angle between the tangent planes along the intersection curves deviates less then 2.5° from orthogonality.

Algorithm 3 Approximate implicitization by approximately orthogonal surfaces

---

Input A parametric space curve $\mathbf{C}$ or a set of sampled points $\mathbf{p}_i$.
Output An approximate implicit representation as the intersection of two approximately orthogonal surfaces.

1: Run Steps 1, 2, 3 of Algorithm 2.
2: Approximate $\|\nabla f\|$ and $\|\nabla g\|$ by polynomials or piecewise polynomials $\bar{f}$ and $\bar{g}$ by minimizing (12.11).
3: Introduce the two auxiliary function F and G as in (12.9) and (12.10), where the norms of the gradients are replaced by their piecewise polynomial approximants.
4: Represent the given curve as the intersection of the two approximately orthogonal algebraic surfaces F, G.

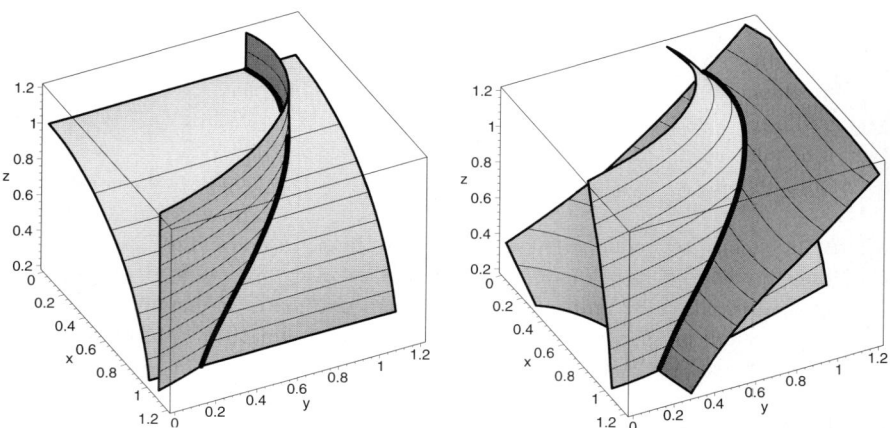

Fig. 12.4. Approximate implicitization of a space curve using Algorithm 2 (left, intersection of two generalized cylinders) and 3 (right, intersection of two approximately orthogonal surfaces).

12.4 Approximate Implicitization of Surfaces of Revolution

A surface of revolution is obtained by rotating a profile curve $\mathbf{q}(v)$ about (e.g.) the z–axis. We propose two techniques for generating an approximate implicit representation by a piecewise polynomial. Both techniques reduce the problem to the implicitization problem of a planar curve.

12.4.1 Implicitization via elimination

First we apply a method for approximate (or exact) implicitization to the profile curve in the rz–plane, where the radius r denotes the distance to the z–axis. For example, one may use the method which was described in Section 12.2. We obtain an implicit representation of the form $f(r, z) = 0$, where f is a (piecewise) polynomial.

In order to obtain an implicit representation of the form $g(x, y, z) = 0$, one could substitute $r = \sqrt{x^2 + y^2}$. However, the resulting scalar field

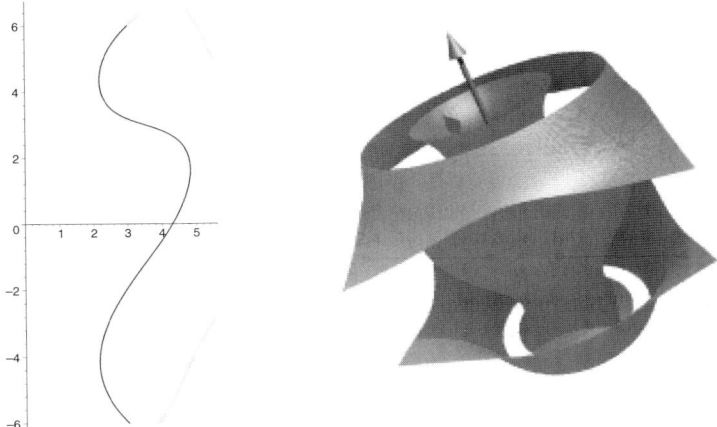

Fig. 12.5. Approximate implicitization of a surface of revolution using elimination, see Example 10. Left: profile curve, right: the surface.

$$(x, y, z) \mapsto f(\sqrt{x^2 + y^2}, z) \qquad (12.13)$$

is no longer given by a piecewise polynomial representation, due to the square root. Instead, we eliminate r using a resultant,

$$g(x, y, z) = \mathrm{Res}_r(f(r, z), r^2 - x^2 - y^2). \qquad (12.14)$$

The degree of g will be twice the degree of f. Clearly, the resultant can be evaluated only if f is a polynomial. In the case of a piecewise polynomial (spline function), this approach has to be applied to the polynomial segments.

Example 10. We apply the technique of Section 12.2 to the profile curve (black line) shown in Figure 12.5 (left) and obtain an approximate implicitization by a bi–quartic tensor–product polynomial (grey curve). After computing the resultant, this leads to an approximate implicit representation of the the corresponding surface of revolution (right). The function g is a tensor–product polynomial in x, y, z of degree $(8,8,8)$. Only even powers of x and y are present. Note that the approximate implicitization produces two additional branches, which do not intersect the surface.

This method for approximate implicitization of surfaces of revolution has two major drawbacks.

- First, in the case of a piecewise polynomial representation $f(r, z) = 0$ of the profile curve, the resulting piecewise polynomial g will not necessarily inherit the smoothness properties of f. E.g., if f is a C^1 spline function, then g will not necessarily be C^1.
- Second, even if the approximate implicitization of the profile curve has no un-wanted branches and singular points in the region of interest, these problems may

be introduced by the eliminating r, see Example 11. Indeed, this elimination is equivalent to computing the polynomial g from

$$g(x, y, z) = f(-\sqrt{x^2 + y^2}, z) \cdot f(\sqrt{x^2 + y^2}, z). \tag{12.15}$$

Note that this produces indead a polynomial, since only even powers of the square root are present! The product (12.15) leads to a *symmetrized version* of the approximate implicitization of the profile curve. Consequently, additional branches from the half–plane $r < 0$ may cause problems.

Example 11. Approximate implicitization of the profile curve (a cubic Bézier curve) by a cubic polynomial using the method described in Section 12.2 produces an implicit curve without additional branches and singular points, see Fig. 12.6, left. However, these problems are present after the elimination step (12.14), see Fig. 12.6, right. The reason for this phenomenon can be seen from the global view (bottom row in the picture): the elimination produces a symmetrized version of the approximate implicitization. Note that methods for exact implicitization of the profile curve have similar problems.

Remark 12. The first problem can be resolved by using Eq. (12.15) instead of (12.14).

12.4.2 Implicitization via substitution

In order to avoid the problems of the first approach, we propose to implicitize the profile curve $\mathbf{q}(v)$ in the rz-plane by the zero contour of a bivariate function $f(r^2, z)$. The bivariate function $f(r^2, z)$ can be chosen from the space of all bivariate functions with even power in r. We may use any basis (e.g., tensor–product B–splines) and express the bivariate function $f(r^2, z)$ as

$$F(r^2, z) = \sum_{i \in \mathcal{I}} c_i \, \varphi_i(r^2, z) \tag{12.16}$$

with real coefficients c_i, where $\mathcal{I}$ is a certain index set. The method for approximate implicitization described in Section 12.2 is applied to this representation. The approximate implicit representation of the surface of revolution is then obtained by a substitution,

$$g(x, y, z) = F(x^2 + y^2, z). \tag{12.17}$$

The degree of g with respect to x and y is twice the degree of F with respect to r^2, while the degrees with respect to z are equal.

Example 13. We apply this approach to the profile curve of Example 11, using a polynomial F of total degree 3. The implicit equation of the profile curve has degree (6,3), and the approximate implicit equation of the surface of revolution has degree (6,3,3). As shown in Fig. 12.7, we may achieve a similar accuracy in the region of interest by using an approximate implicitization of the profile curve that is symmetric with respect to the axis of revolution. Due to this symmetry, no problems with unwanted branches and singular points are present.

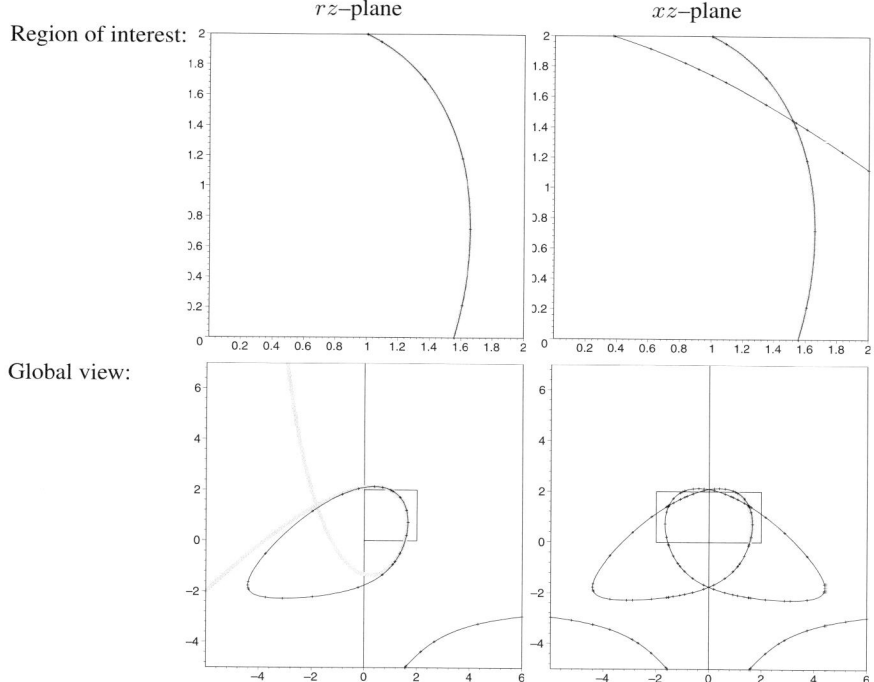

Fig. 12.6. The elimination of r may produce additional branches and singular points. Top row: Region of interest $[0, 2]^2$, Bottom row: global view. Left: Approximate implicitization $\mathcal{Z}(f)$ of the profile curve in the rz–plane. Right: Intersection of the approximate implicitization $\mathcal{Z}(g)$ with the xz–plane. The original profile curve is shown in grey.

Example 14. We consider the discretized profile curve shown in Fig. 12.8, left, and apply the method of Section 12.2 to it. The function F is a bi–quadratic tensor–product spline function whose domain is the union of the cells shown in the figure. This leads to an approximate implicit representation of the profile curve (Fig. 12.8, center) and of the surface (right) of degree $4(\times 4) \times 2$. In the surface case, the spline function is defined with respect to ring–shaped cells, obtained by rotating the cells shown in the left figure.

12.5 Conclusion

Several techniques for approximate implicitization of space curves and surfaces of revolution have been presented. These techniques are based on algorithms for (exact or approximate) implicitization of planar curves. In the case of space curves, a representation of two approximately orthogonal surfaces can be obtained, which provides several advantages, such as a geometrically robust definition of the curve and the

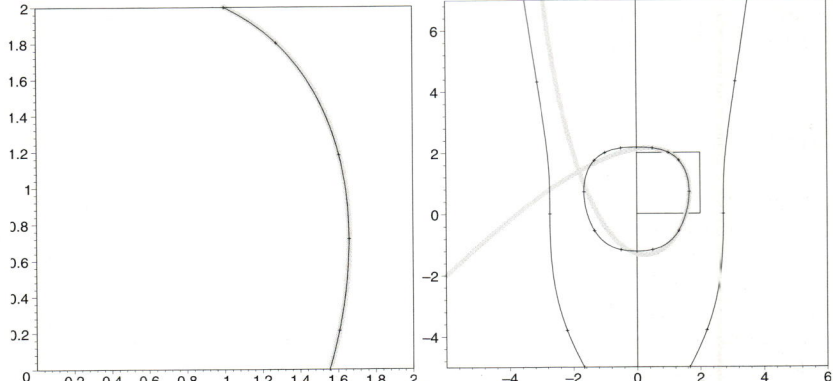

Fig. 12.7. Approximate implicitization of a surface of revolution via substitution avoids potential problems with additional branches and unwanted singular points. Left: Region of interest, right: global view. The original profile curve is shown in grey.

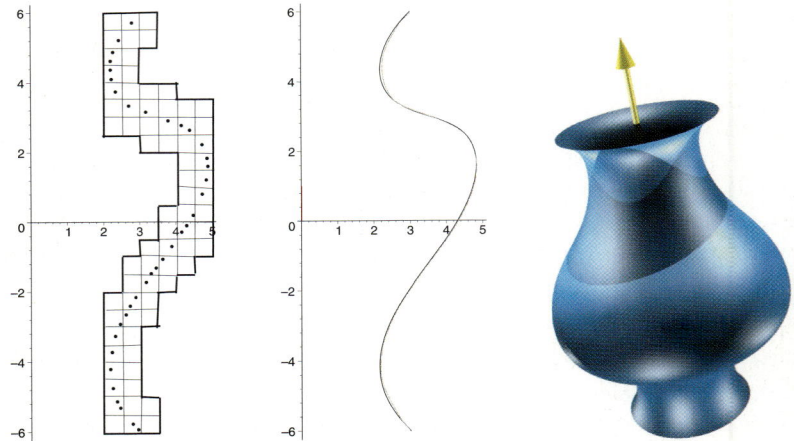

Fig. 12.8. Approximate implicitization of a surface of revolution of degree $4 \times 4 \times 2$, using a biquadratic spline function F, see Example 14.

possibility to obtain a good approximation of the distance field to a space curve. As shown in the case of surfaces of revolution, only approximate implicitization is able to produce a representation which is free of unwanted branches and singular points in the region of interest.

Acknowledgments

This research has been supported by the European Commission through project IST-2001-35512 'Intersection algorithms for geometry based IT-applications using approximate algebraic methods' (GAIA II), and by the Austrian Science Fund (FWF) in the frame of the special research area SFB F013.

References

1. Aigner, M., Jüttler, B., Kim, M.–S.: Analyzing and enhancing the robustness of implicit representations, in: Geometric Modelling and Processing 2004, IEEE Press, 131–140.
2. Chuang, J., Hoffmann, C.: On local implicit approximation and its applications. ACM Trans. Graphics **8**, 4:298–324, (1989)
3. Corless, R., Giesbrecht, M., Kotsireas, I., Watt, S.: Numerical implicitization of parametric hypersurfaces with linear algebra. In: AISC'2000 Proceedings, Springer, LNAI 1930.
4. Cox, D., Little, J., O'Shea, D.: Using algebraic geometry, Springer Verlag, New York 1998.
5. Cox, D., Goldman, R., Zhang, M.: On the validity of implicitization by moving quadrics for rational surfaces with no base points, J. Symbolic Computation, **11**, (1999)
6. Dokken, T., et al.: Intersection algorithms for geometry based IT-applications using approximate algebraic methods, EU project IST–2001–35512 GAIA II, 2002-2005.
7. Dokken, T., and Thomassen, J., Overview of Approximate Implicitization, in: Topics in Algebraic Geometry and Geometric Modeling, AMS Cont. Math. **334** (2003), 169–184.
8. Gonzalez-Vega, L.: Implicitization of parametric curves and surfaces by using multidimensional Newton formulae. J. Symb. Comput. **23** (2-3), 137-151 (1997)
9. Hoschek, J., and Jüttler, B.: Techniques for fair and shape–preserving surface fitting with tensor–product B-splines, in: J.M. Peña (ed.), Shape Preserving Representations in Computer Aided Design, Nova Science Publishers, New York 1999, 163–185.
10. Jüttler, B.: Least-squares fitting of algebraic spline curves via normal vector estimation, in: Cipolla, R., Martin, R.R. (eds.), The Mathematics of Surfaces IX, Springer, London, 263–280, 2000.
11. Jüttler, B., and Felis, A.: Least–squares fitting of algebraic spline surfaces, Adv. Comp. Math. **17** (2002), 135–152.
12. Mourrain, B., Pavone, J.-P., Subdivision methods for solving polynomial equations, Technical Report 5658, INRIA Sophia-Antipolis, 2005.
13. Sampson, P. D. Fitting conic sections to very scattered data: an iterative refinement of the Bookstein algorithm, Computer Graphics and Image Processing **18** (1982), 97–108.
14. Sederberg, T., Chen F.: Implicitization using moving curves and surfaces. Siggraph 1995, **29**, 301–308, (1995)
15. Shalaby, M. F., Thomassen, J. B., Wurm, E. M., Dokken, T., Jüttler, B.: Piecewise approximate implicitization: Experiments using industrial data, in: Algebraic Geometry and Geometric Modeling (Mourrain, B., Elkadi, M., Piene, R., eds.), Springer, in press.
16. Wurm, E., Jüttler, B.: Approximate implicitization via curve fitting, in Kobbelt, L., Schröder, P., Hoppe, H. (eds.), Symposium on Geometry Processing, Eurographics / ACM Siggraph, New York 2003, 240-247.
17. Wurm, E., Thomassen, J., Jüttler, B., Dokken, T.: Comparative Benchmarking of Methods for Approximate Implicitization, in: Neamtu, M., and Lucian, M. (eds.), Geometric Design and Computing, Nashboro Press, Brentwood 2004, 537–548.

Index

A_1 singularity, 47
$\mu^{\mathbb{R}}(d)$, 47
$\mu(d)$, 47

Bézier clipping, 170, 172, 178
Bernstein
  basis, 202
  polynomials, 162, 166
  representation, 170, 171

canal surface, 79
Chmutov's conjecture, 48
circulant matrices, 182, 183
classical invariant theory, 34
classification, 55, 65, 70
computer aided design, 55
conical node, 48
covariant, 31, 38
curve
  parametrization, 129
  rational, 128
  regularity, 204–207
  ridge, 199, 209, 211
  self–intersection, 199, 210, 211
  silhouette, 212
  space, 215
  topology, 199

Descartes' rule, 202
Discrete Fourier Transform, 184

distance bound, 217
divisor, 125
divisor
  anticanonical, 132
Doo–Sabin subdivision, 185
Dupin cyclide, 79–81, 85, 91

eigencubes, 189
eigenpolygon, 184–186
eigenprisms, 193, 194

G–circulant matrices, 181, 187, 188
Gaussian map, 84–86
group characters, 188

Hessian matrix, 60
hypersurface
  isotropic, 81, 84

implicitization, 55
implicitization
  approximate, 216
  via elimination, 222
  via substitution, 224
injectivity criterion, 207
intersection, 100, 101, 104

Jacobian ideal, 60

K3 surface, 48

Laguerre geometry, 80, 84, 92

matrix
    Hessian, 60
Milnor number, 60
monoid, 57, 62, 63, 66, 67
monoid
    hypersurface, 55, 56
    quartic, 55, 64, 71
    real, 62, 75
    surface, 55, 58, 63
multiplicity, 55, 56, 63, 67, 68

natural quadric, 79, 91
node, 47
number
    intersection, 58, 61, 65
    Milnor, 60, 63, 75

orbits, 33
oval, 109

parameterization, 55, 56
parametrization
    curve, 119, 129
plane
    3-projective, 97
    tangent, 97, 100
point
    base, 57, 58
    boundary, 163, 169, 171, 172, 176
    critical, 108, 110
    cuspidal, 174, 176
    pinch, 102, 109
    purple, 142
    ridge, 142
    singular, 201, 204
    solitary, 48
    turning, 163, 171, 172, 174
    umbilical, 142
polygon
    Newton, 119, 125
polynomial
    folding, 48
    real folding, 49
    Tchebychev, 48

quadratic canal surfaces, 81

quadratic spline, 191
quadratically parameterizable surface, 31
quadric
    natural, 79
quartic surface, 31, 48

real line arrangements, 48
representation
    rational univariate, 148
ridge curve, 209
right equivalence, 74
root–system A_2, 49

Sampson distance, 219
scroll, 98
scroll
    dual, 98, 100
silhouette curve, 212
singular locus, 100, 101
singularity, 47
space curve, 215
Steiner surface, 31
Steiner surface
    discriminant, 41
    dual surface, 40
    implicit equation, 39
    triple point, 41
subdivision
    method, 199, 201, 208, 209
    solver, 203
    surfaces, 181, 183, 186
    volumes, 189
surface
    (1,2) bidegree, 96
    canal, 79
    directrix, 95
    intersection, 199, 205
    monoid, 58
    nodal, 50
    normal ruled, 95
    of revolution, 215
    quadratic canal, 79, 83, 85–87, 91
    quartic, 31, 48
    ruled, 95

Steiner, 31
toric, 121
Sylvester resultants, 169

torsal line, 100, 102, 108

Walsh-Hadamard matrices, 189

Printing: Krips bv, Meppel, The Netherlands
Binding: Stürtz, Würzburg, Germany